D1784425

William Leitch Presbyterian Scientist

&

The Concept of Rocket Space Flight 1854-1864

(2nd Edition)

by Robert Godwin

Principal Reverend William Leitch D.D. (ca. 1863)

William Leitch, Presbyterian Scientist & The Concept of Rocket Space Flight (2nd Edition)

Copyright ©2017 Robert Godwin ISBN 978-1926837-40-6

Portions of this book first appeared in 2015 on TheSpaceLibrary.com.

All rights reserved. No part of this book may be used or reproduced in any manner whatsoever without written permission except in the case of brief quotations embodied in critical articles and reviews.

Printed and bound in the USA. Apogee Prime is an imprint of Griffin Media, Burlington, Ontario, Canada, L7R 1Y9

www.apogeeprime.com

Contents

Chapter 6

Chapter 7

Chapter 8

Chapter 9

Chapter 10

Chapter 11

Chapter 12

Second Edition Updates

Introduction

Why write a book about an almost unknown Victorian Presbyterian minister?

In the spring of 2015 during a trip to Seattle, Washington. I came across a copy of *"Space Travel"* by Kenneth Gatland and Anthony Kunesch. Gatland had been a founder of the British Interplanetary Society, when it regrouped after World War II. His book had been written in 1953.

Gatland had come across an unknown author who had very clearly articulated the idea of rocket spaceflight, *obeying Newtonian laws*, in the year 1899. History tells us that Tsiolkovsky had been postulating such things around that time, but he didn't publish until 1903. Robert Goddard had already had his spaceflight revelation while climbing the tree in his family yard, but he was only seventeen years old in 1899 and his idea for rockets was years in the future. The Rumanian/German Hermann Oberth would not publish his ideas for spaceflight until the 1920s and his obscure fellow German pioneer, Hermann Ganswindt who had been talking about rockets in space since 1880, was mostly forgotten. From every known angle it became apparent that this was definitely something worth investigating.

Had Ganswindt inspired someone in the English speaking world? Or had someone spoken to Tsiolkovsky about his theories and brought them back to England at the end of the century?

Gatland's original citation was from an obscure children's volume called *"Half Hours in Air and Sky"* which could be found in Google's book archive. The important quote came from an essay called *"A Journey Through Space"* but the date wasn't 1899, as Gatland had thought, it was from 1877. If you're a space history enthusiast you might appreciate how important a difference this represented.

On digitally thumbing through the book the reader will find that there is no attribution. Therefore, in my case, the next logical thing to do was to find an unusually worded sentence which could be used for a deeper search. On doing this up came *"God's Glory in the Heavens"* by Principal William Leitch D.D., identified as a Canadian University professor. The date on the book was 1867.

This represented an exciting breakthrough as it seemed that this amazing quotation was almost as old as Jules Verne's legendary moon flight story of 1865. But, even then, the book which appeared on Google said "Third Edition" on the title page. Could it really be possible that this quote predated Verne?

Using a subscription database it was possible to find an advertisement for the first edition; in the autumn of 1862. This was the first solid evidence of someone seriously proposing rockets for spaceflight who couldn't possibly have been inspired by Verne (although in this book you'll see how that might have been a fallacious assumption too!)

In the introduction to the book the author mentioned that some of his ideas had appeared in "*Good Words*", an obscure reference which required another trip to Google to discover that it was apparently an early magazine which had first appeared in 1860 in Edinburgh, Scotland. However, there was no online scan of the early volumes. This would require a trip to the Guelph University archives out in the countryside and a retrieval of the bound volumes of *Good Words* from the stacks. Right there in the September 1861 edition was the first appearance of "*A Journey Through Space,*" including the rocket quote.

The next course of action was to contact some of the top authorities on the subject of early rocketry to see if this was as important as it seemed.

Mr. Frank Winter, curator emeritus of rocketry at the Smithsonian has been the most prolific and informed writer in the world on the birth of rocketry. He received the information about Leitch and immediately set about confirming these findings by digging into the Smithsonian archives and his own considerable holdings. The next person contacted was Mr. David Baker in London England, editor of *Spaceflight* Magazine and author of a massive tome of rocket history - the superb volume "*The Rocket*". Finally, Mr. Michael Ciancone, chair of the American Astronautical Society History Committee and the author of an index of books on pre-Sputnik spaceflight, was given the information. It was the conversation with Michael that caused some hesitation. *"How do you know this person wasn't just guessing?"* he asked. It was a good question.

Shortly thereafter Frank and David came back with enthusiastic endorsements and so it was decided that the news should be quietly released. In October 2015 a space conference was being held in North Bay Ontario and it seemed to be a good venue to introduce this new nugget of Canadian spaceflight history. So with the assistance of a colleague in the promotions business, Mr. Hugh Black, we wrote up a press release saying that William Leitch's story would be a part of my presentation to the delegates. What no one expected was that astronaut Chris Hadfield, who was also speaking that weekend, would "Tweet" the press release to his massive global audience. Within 48 hours the story was in half a dozen languages on thousands of news sites from the Daily Mail in London to the Associated Press in the United States. People in China and India were reporting the story. Three days later it was on page three of the biggest national newspaper in Canada and the phone was ringing for television interviews.

But despite all of this fuss, Michael Ciancone's challenge was still hanging in the air. Could this have just been a lucky guess by Leitch; a wild proclamation similar to that of the 17th century lunatic Cyrano de Bergerac? It was Michael's challenge that led to the book you are holding. How much proof could be accumulated to prove that Leitch was imbued with the knowledge to sensibly make his assertion about rockets?

Digging into the mysteries of the Reverend William Leitch took the better part of a year. He hailed from Scotland and spent most of his life in the county of Fife where my own ancestors had lived. One newspaper that was uncovered had Leitch's name just one column over from my own Gt. Gt. Grandfather's name. They must have known each other. Both were granted official biographies in the "Men Of Fife" series and both spent a lot of time in the tiny village of Cupar in the 1850s, Leitch as the moderator of the Presbytery and my ancestor as one of the local Provosts. A nice coincidence to be sure, but it was the other astonishing connections that Leitch had with the Victorian science community and the big science and political questions of his time which leads me to emphatically answer Michael's question with a resounding "This was not a guess." After reading this book, I hope you will agree with me.

Robert Godwin (Burlington Ontario 2016)

Since the first publication of this book a substantial amount of new information about William Leitch has come to light. It is included here as an entirely new section beginning on page 142.

Robert Godwin (Burlington Ontario 2017)

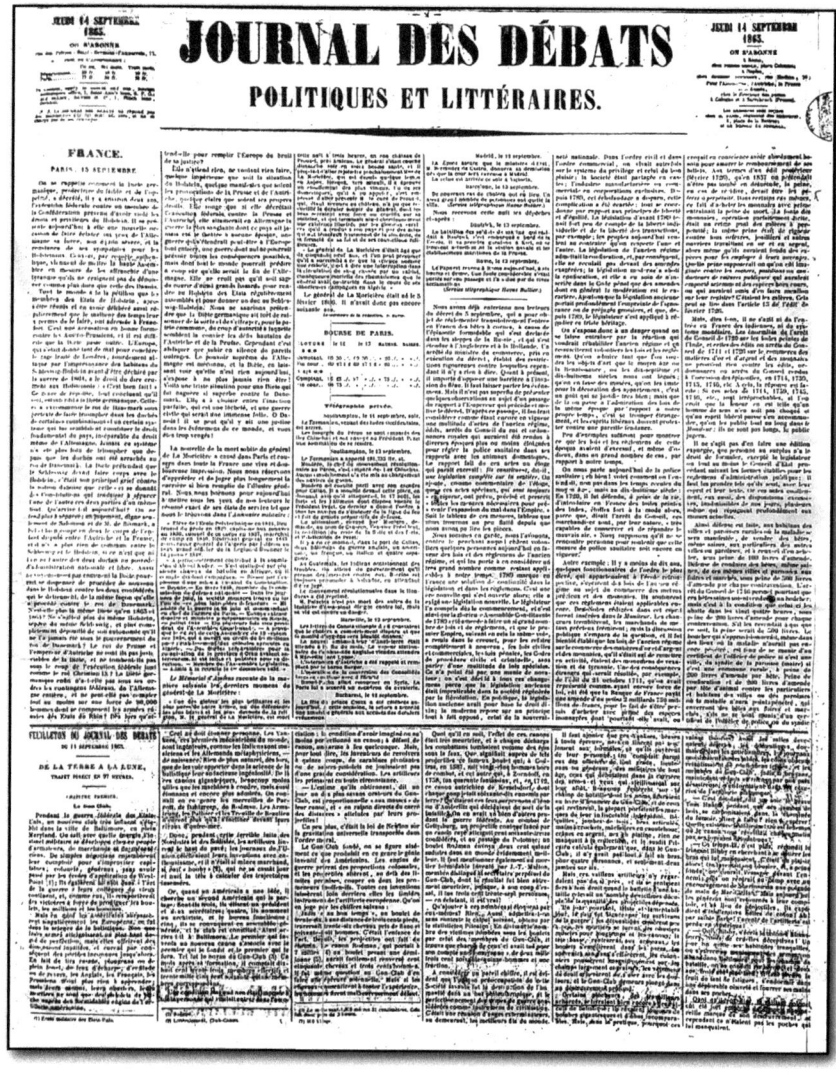

The very first appearance of Jules Verne's De La Terre à La Lune - September 11 1865

Chapter 1

If history text books tell us anything, it is to question things and to never rest on our laurels and always be prepared for change. Indeed it is with the invention of history that we first became consciously aware of change. The entire prevailing history of rocketry tells us that modern space travel was the invention of a few singular minds, most of whom were working in almost solitary conditions. Those names now ring down to us through history texts spanning more than a century of scholarly achievement. But who were these men, and who inspired them? The most frequently cited source of inspiration was a legendary French writer of fiction.

Jules Verne

On February 8th 1828, Pierre Verne and his wife Sophie Alotte de la Füye, announced the birth of their first child, a son named Jules. In 1841, when he was only thirteen years old, the daring young Verne decided to participate in the new era of adventure by signing on as a cabin boy aboard a passing merchant ship. Needless to say his father knew nothing of this until it was almost too late. Pierre quickly found out where his son had gone and intercepted him at the next port. Fortunately the future of space history was spared by Jules' capture. As you will see in the pages ahead, Jules Verne was not the only young teen attracted by the romance of adventures at sea.

In 1848 France was in the throes of revolution and Jules Verne found himself spending much of his time in and around the Parisian artistic community. The revolution had spawned hundreds of new publications and good writers were in demand. He enjoyed writing and his artistic friends told him that he had potential. As Verne gradually delved further into this new occupation it became apparent to him that he had found his place.

Verne would spend many hours at a café on the *Lycée Henry IV* discussing ballistics with his cousin Henri Garcet, who was a mathematician and also an author of books on science and mechanics. The result of these conversations would be published by Verne as a serial, beginning on September 14th 1865 and then running for a month, in the magazine *Journal des Débats*. Verne's story came in two parts known as *From the Earth to the Moon* and *A Trip Round the Moon*. The latter would not have its first publication until over four years later when it also appeared in the same magazine. After serialisation, both would be published as books.

Jules Verne (ca. 1860)

Verne's lunar story begins in Baltimore, Maryland, and concerns the exploits of a group of American sportsmen who decide to alleviate their boredom by building the world's biggest gun. The purpose of the enormous enterprise is to use the weapon to send a shell to the moon. Verne studied ballistics (and conversed extensively with his cousin) and then drew up the plans for a gun large enough to launch a projectile to the moon. Although Verne's cousin Garcet had helped with the mathematics, it was Verne who did the calculations for the legendary spacecraft.

Undoubtedly one of the more remarkable elements of the story is Verne's choice to construct the gun in Florida at a fictional location only a few miles from where the Kennedy Space Center would stand a hundred years later. It is very likely that Verne understood the dire implications of putting a human inside his projectile and there is little doubt that *From the Earth to the Moon* was intended to be satirical.

The gun is dutifully fired and the daring crew of three, a Frenchman and two Americans, is sent on its way. What transpires next is in the second book entitled *Round The Moon* (more often than not, the two are thought of as one book). It is quite ironic since ultimately they do not reach the moon, but instead have to settle for circumnavigating the satellite and then returning back to Earth. Verne went to great lengths to spell out the details of the flight, even though most of the mathematics would have been entirely wasted on many of his readers, before he brought his valiant crew back to a splashdown in the Pacific Ocean. This story was read by a young man in Russia.

Konstantin Tsiolkovsky

On September 17th 1857 in a small village called Izhevskoye in Russia, a boy named Konstantin was born to the Tsiolkovsky family. The young Tsiolkovsky loved to read and he remembered having a happy childhood. At the age of nine he contracted scarlet fever which led to deafness; consequently he was unable to attend school and so he was sequestered from the world at large. By the age of fourteen he had found some solace in reading and fortunately his father owned many books on mathematics and science. Tsiolkovsky would become a voracious reader, consuming anything he could find, especially the sciences. His father encouraged him and sent him off to further his education in Moscow, where he barely survived; living on a diet of brown bread and water for days at a time.

At the age of 19 Tsiolkovsky moved back home from Moscow and by the time he was 21 he took up a teaching position at the local school in Kaluga where he taught mathematics and geometry.

While he lived on his meager teacher's salary he established himself with something of a reputation as the archetypal "mad scientist." Experiments abounded in his apartment. By 1883, Tsiolkovsky wrote *Free Space*, a book in which he had turned his attention to Newtonian physics. He was interested in how objects were affected by outside forces in a neutral environment; such as space.

Most significantly he realized that Newton's law of conservation of energy would be *more* important in space. Newton's *first* law stated; Every particle continues in a state of rest, or a state of motion with constant speed in a straight line, unless compelled by a force to change that state. The *second* law stated; the net unbalanced force producing any change of motion is equal to the product of the mass and the acceleration of the particle; and finally the *third* law stated; that all forces arise from the mutual interaction of particles, and in every such interaction the force exerted on the one particle by the second is equal and opposite to the force exerted by the second on the first.

What the final law actually meant in plain English is that for every *action* there is an equal and opposite *reaction*. Perhaps the single most important fact that Tsiolkovsky established in his long and industrious career was the realization that a rocket would work in the vacuum of space.

On March 28th 1883 he wrote, *"Consider a cask filled with a highly compressed gas. If we open one of its taps the gas will escape through it in a continuous flow, the elasticity of the gas pushing its particles into space will also continuously push the cask itself. The result will be a continuous motion of the cask."* He also postulated that if the cask was to have multiple taps they could be regulated in such a way to make it possible to actually steer the cask.

Now this may not seem particularly startling today, but it is worth noting that as much as 50 years later some people still couldn't accept the fact that a rocket didn't need something to *push against*.

Konstantin Tsiolkovsky (ca. 1880)

The absolutely crucial element here is that Tsiolkovsky knew immediately that a rocket would work in space in exactly the same way as a child's balloon splutters its way around a room when released - by internal reaction.

Tsiolkovsky spent much of his time working on all aspects of aerodynamics but on August 25th 1898 he would put pen to paper with his design for a rocket. He claimed he was inspired by a book published just two years earlier called *New Principle of Aeronautics*, written by A.P. Fyodorov, in which a jet-propelled vehicle is described. Or perhaps he heard something of a work by I.V. Meshchersky published in 1897, called *The Dynamics of a Variable Mass Point*. In this book the author presented an equation that described the motion of an object with variable or changing mass; even citing a rocket as an example. However, Tsiolkovsky would later write, *"For a long time I thought of the rocket as everybody else did – just as a means of diversion and of petty everyday uses. I do not remember exactly what prompted me to make calculations of its motion. Probably the first seeds of the idea were sown by that great fantastic author Jules Verne: he directed my thoughts along certain channels, then came a desire, and after that, the theory. The old sheet of paper with the final formulae of a rocket device bears the date August 25th 1898. I have never claimed to have solved the problem in full."*

But what about Alexander Petrovich Fyodorov? Thanks to Tsiolkovsky's achievements he would get some credit in his later years. Tsiolkovsky referred to the obscure booklet written by Fyodorov as being like Newton's apple and the factor which had crystallized the idea in his mind to use a reaction engine in spaceflight. Fyodorov postulated that such an engine could be powered by liquid fuels (carbon disulphide and nitric acid) and he requested that anyone who agreed with his idea should contact him if they wanted to finance the construction of his rocket. Fyodorov would go on to become a populariser of science in magazines such as *Science and Life* and would even create his own magazine called *Polytechnica*.

The first publication of Tsiolkovsky's ideas for rockets would not come until 1903, the same year as the Wright brothers' legendary first flight of a manned, heavier-than-air, powered device. His article appeared in the May issue of the magazine *Science Review* and was called *Investigating Space with Rocket Devices*. Tsiolkovsky noted that in a rocket motor Newton's third law would apply. However, what he also realized was that the Newtonian laws of mechanics were based around a system where the mass of the object was fixed. Tsiolkovsky asked, *"What happens if the mass changes?"* His answer to this basic question was what made space travel theoretically possible.

But someone had to build the hardware.

Robert Goddard (ca. 1925)

Robert Goddard

Robert Hutchings Goddard was born on October 5th 1882 into a respectable middle class family in Worcester Massachusetts. From a very early age he showed a keen intellect and an intensely inquisitive mind. He used these talents to investigate different aspects of the burgeoning new technology spawned by the Industrial Revolution. On October 19th 1899 Goddard braced a small wooden ladder against the cherry tree in his yard and climbed into the branches. He had just passed his 17th birthday and he was daydreaming when, he later recollected, that he had a vision. Not a hallucination or numinous visitation from God, but a vision of the future, a future which he was convinced could exist. Goddard dreamed that he went to Mars and he arrived there using a machine of his own design. He later climbed down the tree and went inside, whereupon he made a notation in his journal. Almost every year for the rest of his life he would remember and celebrate the anniversary of that day. Like Tsiolkovsky, Goddard had also been an avid reader of science fiction; including the works of Jules Verne.

On June 15th 1911 he graduated *cum laude* from his class at Clark University and the next day he was introduced for the first time as Dr. Goddard. Later that summer he read an article in the *Boston American* entitled *"When May We Go to the Moon?"* He continued to be encouraged by his peers, who urged him to direct his obvious talents toward a suitable research project. On the anniversary of his life-changing dream he wrote in his diary that he was looking at various different means of propulsion.

On June 2nd 1913 he started writing the specifications for his first rocket patent. On October 1st the patent was filed at the US Patent office and then was subsequently issued on July 7th of the following year. The accompanying diagrams showed a two stage rocket system using cylindrical discs of solid fuel, an elongated tapered exhaust nozzle, a gyroscope for stabilization and small solid fuel axial thrusters that would be used to *spin the rocket*. The text of the patent stated, *"This invention relates to a rocket apparatus and particularly to a form of such apparatus adapted to transport photographic or other recording instruments to extreme heights."*

Between February 28th and March 11th of 1914, frail and bed ridden with a debilitating disease, Goddard drew up his second rocket patent. After mailing it to the patent office he spent two weeks rethinking the problem. On March 26th he made some modifications, this time the patent would include not only the design for a shotgun-like multiple cartridge launcher but also for a liquid fuel rocket. This was a monumentally important step forward for Goddard and would ultimately assure his immortality in the ranks of space scientists. In late 1919, the Smithsonian published Goddard's small booklet entitled, *A Method of Reaching Extreme Altitudes* in which he outlined all of his principals for building an efficient rocket—although a solid-propellant, unmanned type— for launching into space.

On January 12th 1920 the *New York Times* published a front-page editorial chastising Goddard for not understanding that a rocket couldn't work in space because it had no air to *push against*. This ill-informed article embarrassed the notoriously retiring Goddard into working in ever-tightened secrecy. After another six years of quiet research he finally built and launched the first liquid-propelled rocket from a field in Massachusetts. It was March 16th 1926. He later expended considerable effort proving experimentally that a rocket *would* work in a vacuum, *and would do so more efficiently than in the atmosphere*. He built an ingenious tube-shaped vacuum chamber for this purpose. Decades later the *Times* would publish an apology to Goddard as men flew to the moon aboard Apollo 11.

Hermann Oberth (ca. 1920)

Hermann Oberth

In 1914, while Goddard was haggling with the US military over money to build rockets, a young man named Hermann Oberth was being drafted into the Austro-Hungarian army. Oberth was 20 years old and he had spent most of the previous few years convalescing from a serious bout of scarlet fever. Oberth was born in Transylvania on June 25th 1894, his father was a popular family doctor. When the First World War began Hermann was forced to fight as an infantry-man.

Oberth had also been pondering the problem of space travel since reading Verne's *From The Earth to the Moon* in the winter of 1905-6. The great fiction writer had inspired the eleven-year-old Oberth to consider the notion that leaving the Earth was actually possible. Hermann Oberth maintained that at a very early age he had independently calculated the acceleration necessary for a successful launch into space. In an interview later in life he said that another science fiction writer named Hans Dominik who combined a space gun with anti-gravity made him begin to work on a better solution.

"A Trip to Mars" by Hans Dominik (1909) caused Oberth to think about better solutions than a gun.

Oberth realized that there would be no feasible way to use a gun for the launch, as Jules Verne's had suggested, and so he turned to rockets. The problem with rockets was that it seemed impossible to create an exhaust velocity sufficient to lift the weight of the rocket into space. In 1909, after reading Dominik, Oberth drew up his first plans for a rocket. It was a solid fuel rocket but it was, nonetheless, a radical departure from the Vernian cannon. It would not be long before Oberth had evolved his theory away from solid fuels and turned to the much higher energy potential of liquid propellants. He claims that he first designed a rocket using hydrogen and oxygen in 1912. It would not be until the summer of 1930 before Oberth would finally build his first primitive liquid fuelled rocket. His disciple was the young son of a German baron named von Braun.

Wernher von Braun, born in 1912, and destined to become the most famous rocket maker in history also knew, thanks to Oberth, that the rocket was the best solution for space travel. It would be his efforts and those of his Soviet counterpart Sergei Korolev that would finally send humans into space. Both also acknowledged Jules Verne as their inspiration.

Hermann Ganswindt (ca. 1900)

Hermann Ganswindt

The one man who gets almost no credit, because history tells us that he was generally unlikable and wildly eccentric, was Hermann Ganswindt who claimed to have suggested a reaction propelled spacecraft in 1880. His vehicle used a continuous stream of explosives to push its way through space. It is stated by writer Willy Ley in his history of rocketry that the earliest date he could confirm that Ganswindt had first postulated these ideas was around 1891.

"In the vacuum of space of course you cannot catch the air with wings and glide down. So what must one do in order to still overcome gravity and rise up? Answer: - - you simply take with you a mass of air in the form of explosives, which also contains in itself the highest form of power! Thus it can be construed as a flying machine based on the reaction laws of exploding substances. I've invented this kind of flying machine rather than a flying machine with wings. Exact calculations have shown, however, that such a flying machine can only be driven with explosives in very economical terms of power consumption when it assumes a very high travel speed so that it

*Hermann
Ganswindt's
rocket spacecraft
(ca. 1899)*

*is not suited for use here on earth, because the resistance of the air
at such a tremendous speed becomes an obstacle. But the situation
is different in airless space where there is no impediment to even the
speed of a meteor or even that of a comet. And just such a speed is
what we need for an expedition through the universe, because with
the great distance of the heavenly bodies from each other a snail's
pace would not get us to the goal. If for the purpose of the expedition
one's eye catches, e.g., the planet Mars (because the moon would
be uninhabitable) it is a very similar world to our earth, and is the
nearest to us, but the distance amounts to the small sum of 8 million
miles."*

Gandwindt's concept completes our brief précis of the history of
the invention of the rocket for space flight. We have the whole neat
and tidy story of how a French science fiction writer proposed using
a gun for a manned scientific trip to the moon. How his intellectual
heirs, beginning in the late 19th century turned to the gradual and
more viable proposal of using the rocket, of how they realized that it
would work better in the vacuum of space; how they were subjected
to ridicule for making that suggestion, and how they went about
proving that this was true.

The fundamental discovery which led to spaceflight was the notion that a rocket, a device driven by Newton's law of reaction,
would be the only thing suitable to push a vehicle through space and
that it would in fact work better in space than in the atmosphere.

The men credited with this fundamental truth were first ridiculed,
then honored, and finally acclaimed for their work. But the play is
missing a player. There was another man who was aware of this basic truth long before Ganswindt, Oberth, Goddard, Tsiolkovsky, and
yes, even before Jules Verne. His name was William Leitch.

Chapter 2

Early Life – Rothesay, Greenock – 1814 - 1830

William Leitch was born on May 20th 1814 in Rothesay, on the Isle of Bute, an idyllic outpost located near Glasgow off the west coast of Scotland. It is separated from the mainland by the Firth of the River Clyde and the Kyles of Bute a spectacular carved waterway often seen on Scottish postcards. Ferries have been regularly traversing the 10 km width of the Clyde from Bute to the mainland for generations. The Leitchs were highly visible citizens in the port town of Rothesay, the main town on Bute, and before that in the small villages and towns of mainland Argyle.

William's parents were John Leitch, a customs inspector in the town of Rothesay, and Margaret Sharp. William's grandfather was also named John and he was part of the anti-burgher movement preaching in the tiny hamlet of Toward. On a clear day you could stand on the doorstep of the church in Toward and see across the 5 km of water to Rothesay. The history of the Scottish church is a complex and convoluted story which would require many pages to explain, but it is sufficient to say here that the anti-burgher movement to which John Leitch (Sr.) belonged was one of the very first in the Christian church to refuse to allow secular politics to interfere with their religious practices. Ultimately the position of the anti-burghers can be seen as the early stirrings of the separation of church and state.

One of the major schisms within the Scottish church occurred in 1761 leading to the secession of many ministers in the outlying regions of Scotland, and so by March of 1767 the people of Rothesay were in desperate need of a minister who could preach to them in Irish. John Leitch Sr. was one of the secessionists who was approached to fill the vacancy in Rothesay and after agreeing to take the job he built a house which he called "The Claddy" on the corner of Bridge Street and Bridge End Street, which at that time was still on the waterfront. On 11 March 1778 John Leitch gained permission to build a new church in Rothesay.

William's father, John Leitch II, also grew up to become a respected citizen of Rothesay and his six children were never far from the waterfront where a constant fleet of shipping could be seen plying the Clyde in and out of Glasgow. John Leitch III also became a well-known scholar who socialized with some of the brightest men in Britain and translated scientific works into English.[1]

In 1828 at the age of 14 it seems young William Leitch was in pursuit of adventures at sea (similar to the later infatuations of his illustrious contemporary Jules Verne) when he fell from the mast of one of the many sailing ships in Rothesay harbour. He fractured his hip with an impact so severe it wasn't clear that he would survive. He would remain infirm for the rest of his life.[2]

Condemned to a long convalescence the only solace he could find was in reading books. On weekdays he chose to read about mathematics and science, but on Sundays the only acceptable reading material was scripture, and so for more than a year the impressionable teenager studied and struggled to better understand the universe from the confines of his bed; bouncing back and forth between two seemingly irreconcilable interpretations of the truth. The parallels between Leitch's early circumstances and those of his more famous intellectual heirs are remarkable.

Rothesay Harbour (ca. 1890)

As we have seen, the three men given the most credit for recognizing rockets as a means to escape the confines of the Earth are the Russian Konstantin Tsiolkovsky, the American Robert Goddard and the German/Rumanian Hermann Oberth. All three men were struck down in their youth by an assortment of ailments; Tsiolkovsky with Scarlet Fever, Goddard with Tuberculosis and Oberth with Spanish flu. In all three cases the budding genius was bed-ridden for a lengthy convalescence and found solace in reading; specifically books that contemplated bigger things. So it was with William Leitch.

Once he had regained his mobility William was sent for his primary education at the Parish School of his native town, and then completed his preparatory studies at the Grammar School of Greenock, another waterfront town on the Firth of Clyde, where he studied under the superintendence of the distinguished scholar and expert in botany and fossil collecting, Dr. Thomas Brown. This instilled an interest in Leitch for biology and botany.

The most distinguished alumni of the small school in Greenock was one of the most famed engineers in history, the man who had fixed Newcomen's steam engine and brought the world the industrial revolution, James Watt. William's later writings would suggest that during his studies he also became aware of the works of Laplace, Newton, Descartes, Darwin, Wollaston, Le Verrier, Brewster, Young, Davy, Whewell, Herschel, Hansen and many others.

Around this time William Leitch's brother John was living on Great Russell Street in London and attending King's College (which was much later famous as the alma mater of Sir Arthur C. Clarke, and also where the double helix of DNA was first documented, amongst many other notable achievements.) John Leitch acquainted himself with many of the smartest men in the fields of science and theology and took a particular interest in the cross-over between religion, mythology and science in ancient cultures.

Many years later John Leitch translated, from the German, a book by Karl Müller entitled "*A Scientific System of Mythology*", William was a "subscriber" to his brother's publication. The book was hailed as an important text in the understanding of the correlation between ancient societies and their mythological and religious creations. This subject would also fall close to William Leitch's heart because it represented a reconciliation between science and myth, as well as religion. This may have contributed to the extremely enlightened posture he carried with him for the rest of his life.

The priesthood had skipped a generation in the Leitch family. While his father attended to his duties as a customs officer young William was the next to be groomed as a minister. At the age of sixteen he was matriculated to the prestigious university in Glasgow, which at that time was one of the most celebrated centers of learning in the world. James Watt had worked there a generation earlier (but not as a student or faculty member.) One of the tasks Watt had undertaken was to maintain and repair instruments, and most particularly the optical instruments bequeathed to the University in 1755 by Alexander Macfarlane for conducting astronomy. By the time Leitch arrived, James Watt was long gone, but another prodigy was about to arrive on the scene, a young man named William Thomson, who was one day destined to become Lord Kelvin, the most famous scientist in the world. Thomson's father James became professor of Mathematics at Glasgow in 1832 and taught Leitch. Professor Thomson's wife had died in 1830 and he had been left alone to raise a substantial brood of children. In one respect this seems to have worked in the young William Thomson's favour as he was allowed to sit in on his father's classes, the same classes attended by William Leitch. Students generally ranged from the ages of 14 to 24 but the young Thomson matriculated to Glasgow in October 1834 at the age of ten years old when Leitch was working on his BA.

Glasgow University 1831-1843

Leitch had entered the University of Glasgow in 1831 and for the next three years he studied Greek, Logic and Ethics. In 1837 he received his B.A. before following it up with an M.A. in 1838. He studied astronomy under the tutelage of John Pringle Nichol.[3] Another student in Nichol's class at that time was the aforementioned William Thomson who in May 1836 had also won a prize translating one of the works of Lucian of Samosata; which is of particular interest because it emphatically places the works of Lucian at Glasgow, and probably in the same room as Leitch. Today Lucian is universally accepted as the earliest writer of a fictional trip to the moon. In one of his "Menippean" dialogues, written in the second century A.D. titled *Icaromennipus*, Lucian took his readers on a deliberately satirical voyage into space using the wings of an eagle and a vulture. So it would seem that an education at Glasgow in the early 19th century can be seen to have covered a wide range of disciplines, including what is considered by many to be proto-science fiction.

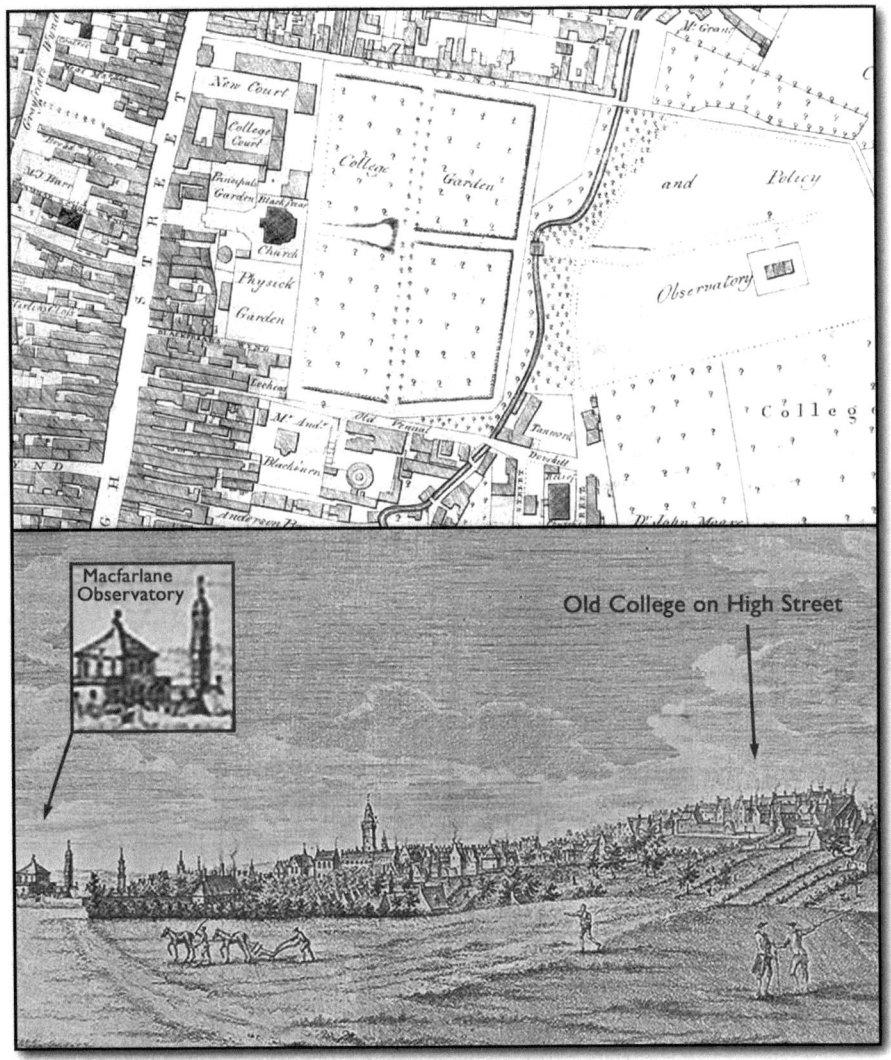

Macfarlane on 1776 map and 1762 engraving

In 1757 the university set up an observatory for Macfarlane's equipment, which included a ten foot long, ten inch reflecting telescope built by William Herschel. The site selected was located on the other side of a fresh water stream called the Molendinar Burn.[4] It was a brisk 400 yards walk across a ten acre college garden, south east of the College Court on High Street. The area was referred to as Dovehill but earlier maps listed it as *Dowhill*. This has led to many historians confusing it with the later observatory erected for John Pringle Nichol at *Dowanhill*, which is three miles north west of Dovehill. Leitch and Thomson both studied at the Macfarlane in the late 1830s. It looks like the two may have competed for prizes in class. Leitch evidently had some skill with engineering subjects winning in 1837 *"For the best account of the recent improvements and applications of the Steam Engine, as a propelling power."* He won again in 1838 for his essay *"On the Construction, Methods of Adjustment and Formulae for the Correction of the Transit Instrument, accompanied and illustrated by a List of Actual Observations made by the writer in the Observatory"*. Thomson and his brother James were trained to use the very same instrument by Nichol. After Leitch graduated Thomson won the prize in 1839 for his astronomy essay *"On the Figure of the Earth"*

Leitch's talents earned him an offer from Nichol to work at the observatory. He had apparently shown a good enough grasp of mathematics and astronomy for Nichol to award him the thankless job of assistant observer. Leitch worked long nights with Nichol learning the latest secrets of the astronomer's art and became aware of the mechanics of our solar system.[5] By the time Leitch worked at the Macfarlane Observatory it was already surrounded by factories and was prone to flooding, making it almost unusable, so a location on Horselet Hill near Dowanhill on the western outskirts of Glasgow was selected for a new bigger and better observatory. (The Macfarlane would be deliberately demolished in 1856 to make way for a railway. Today a National Health building for the homeless sits near the location.) After much discussion and fund-raising, beginning in 1835, the new building was approved, but with the specific intention that it have a magnetic observatory as well as the traditional optical observatory, and that it would be able to make extra-meridional observations. This was a time when the Royal Society was becoming keenly aware of the importance of magnetic observations and it was almost exactly at this time that the Royal Society proposed building similar observatories elsewhere, including in Toronto, Canada where Leitch would ultimately become a university senator. The new Glasgow observatory began construction on the summit of Horselet Hill in 1840 and the Nichol family took up residence in early 1841. That same spring William Thomson left Glasgow without ever taking his degree.

The new college observatory was erected on the highest piece of ground in the area and thus was vulnerable to bad weather. In March 1845 the magnetic building was completely destroyed by gale force winds and the enormous telescope was ripped right out of the ground.

The Horselet Hill observatory (ca. 1841)

The observatory at Horselet Hill is also now long gone, replaced by Notre Dame High School. The only clues to its existence are today's Horselethill and Observatory Roads. It was after graduating in 1838 that Leitch assumed the position of assistant to Nichol.[6] He worked at Dovehill but it is not clear if he ever worked at Horselet Hill. However, Nichol's son, who as a young boy actually lived at Horselet Hill, meticulously described Leitch's working habits and Leitch himself wrote about the facility in great detail, implying that he might have briefly worked there.[7]

In 1841, when the observatory first opened, the only thing in close proximity was the Royal Botanical Garden, which stretched from the gates of the observatory down to the River Kelvin. These gardens were opened with great fanfare in 1842. We can perhaps presume that later visits also contributed to Leitch's interest in Botany; a passion he later passed on to his son.

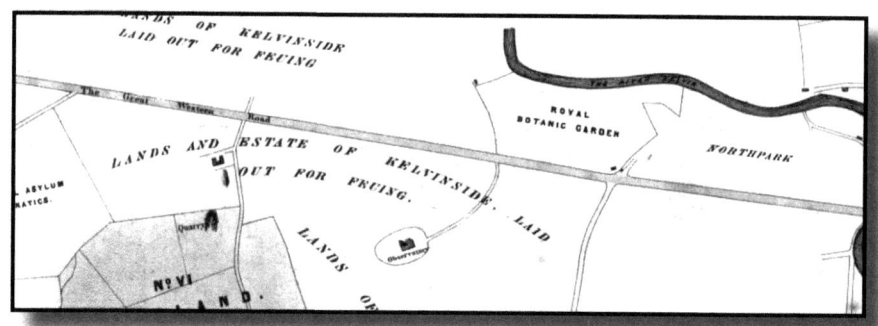

Map of the Observatory & Botanical Gardens

Nichol had taken on the role of Professor of Astronomy in 1836 but at that time astronomy was an awkward fit for the curriculum. No one was teaching the mathematics necessary for a true astronomy course and so the teaching had been erratic at best. One of Nichol's predecessors for teaching astronomy was William Meikleham who had since moved on to teaching the natural philosophy class. Leitch had come in first in Meikleham's class *"for propriety of conduct, exemplary diligence and display of eminent abilities in Examinations on the subjects of the Lectures, and in Essays and Investigations in Physical Science"*. Meikleham was rapidly approaching retirement, and when he began missing his science classes due to ill health, Nichol took over for him. Despite being well-liked by his students, Nichol had something of an unsteady reputation with the school administrators due to a penchant for spending large sums of money on equipment and upgrades for his observatories. It seems he was also frequently on the move, travelling to North America and at least once on a vacation with the Thomson family. It could have been during his trips abroad when Meikleham was sick that Leitch took on the temporary role of stand-in teacher. Not completely new to the job, Leitch had already earned some experience lecturing on mathematics at the prestigious Anderson Institution, which was named after its genius founder who had excelled in the science of ballistics.[8]

In 1838-39, Leitch taught in both Meikleham's science class and Nichol's astronomy class.[9][10] It cannot be stated with any certainty that Leitch actually taught the young William Thomson, who was taking both classes, since Thomson's notebooks at Cambridge don't mention who was at the front of the class on any given day, but the two must have crossed paths. William Thomson, was there for three years after Leitch earned his MA, so it seems likely that Leitch may have been his part-time instructor. In his memoirs Thomson referred to this time of his life as a "white era" because the instruction was so illuminating and by 1849, long after Leitch's departure, he would take over Meikleham's post at Glasgow and would remain there until his death. [11]

Norman Macleod (ca. 1860)

Leitch's friends at Glasgow included landscape painter Horatio McCulloch, the Duke of Argyle, and most importantly Norman Macleod, an up-and-coming minister. This group, along with Leitch's brother John and several others, established an informal club which met regularly for *"the interchange of wit, and of literary productions, whose chief merit was their absurdity."* In 1836 they published a satirical pamphlet titled, *"Sparks of Promethean Fire; or Chips from the Thunderbolts of Jove."* Two of the poems contained therein were written by Macleod to honour his friend William Leitch, whose nickname was "Boss". They were *"Professor Boss's Drinking-Song,"* and *"Invocation to Professor Boss, who fell into the Crater of Hecla."* Both can be found in Macleod's memoirs and involve volcanoes, whales, Greek and Viking gods and many other amusing characters. [12]

Macleod would become an important force in Leitch's life in the years ahead, but circumstances would separate them after completing their studies at Glasgow. They would each independently begin rapid ascents in the hierarchy of the Church of Scotland. Macleod would soon become tutor to the children of an important English family. In this position he found himself introduced at the Weimar court before returning to Scotland to become one of Queen Victoria's ministers at Balmoral, eventually becoming one of her most trusted confidants. Macleod was also acquainted with Wordsworth, but he considered Leitch to be his best friend.[13]

Leitch was a religious man, as were most of his peers, but he was struggling to reconcile the strict interpretations of scripture with what he could see through the telescope on a clear night. Before leaving Glasgow a group of students had gathered together enough money to purchase a telescope for Leitch as a gift. This was a 6½" Gregorian reflector manufactured by James Shortt. Years later visitors to his Manse marveled at the enormous instrument sitting in the main hallway. It would later be donated to Queen's College in Canada. A description of his home later stated, *"The gigantic telescope in his lobby, which his Monimail parishioners contemplated and spoke of with awe; his microscopes and other apparatus filling his study; the last scientific journal on his table; all gave indication that his was a mind that loved to keep abreast of the science of the day. Even when he took to keeping bees, the thing was done not as a recreation, but as a matter of science."* [14][15]

Leitch had of course studied Divinity at Glasgow, coming in joint second in his class in 1836, and in 1838, he completed the four year curriculum to be licensed as a preacher. On leaving university at the end of 1839 he became assistant to the Minister of Arbroath.[16] From there he went to Kirkden before being head-hunted by the 8th Earl of Leven, an up-and-coming figure in the Royal Navy. The Earl proposed Leitch fill the role of Minister of Monimail, a parish situated south of Dundee in the presbytery of Cupar in Fife.[17]

On the 8th of September 1843 the Cupar Presbytery Court in Fife ordained the 29 year old Leitch as minister of Monimail. After the ceremony the group gathered for a "very happy afternoon" at Gardener's Hall in Letham, a few hundred yards from Monimail church, where they "partook freely of sundry of the good things of this life".

Leitch settled into the role of country minister and within three years he had found his future bride, Euphemia Paterson, living a short two mile walk away in Cunnoquhie.

Monimail Ministry 1843 - 1852

Within half a year Leitch had endeared himself to the local parishioners who bestowed a silk pulpit gown on him. Evidently the Countess of Leven had begun a subscription amongst the local populace to pay for the expensive gift. Leitch became firm friends with the Countess who was the daughter of Sir Archibald Campbell, the 2nd Baronet of Succoth and a Fellow of the Royal Society of Edinburgh. In the summer of that year Leitch, Campbell and the Countess along with her husband David Leslie-Melville, 8th Earl of Leven invited children to the ancient home of the Levens, Melville House, to watch as the students of the local girl's schools were tested on their various study subjects by Leitch. This tradition seems to have continued for several years, with the children usually sitting on the lawn of the old mansion and subsequently given gifts and nourishment.

Another role which Leitch had to assume as the local minister was the hiring of teachers for the parish. He also wasn't above leaning on his contact with his friend Norman Macleod to find a specialist preacher or two when the workers on the local railway were in need of guidance in Gaelic. The construction of railways all over the country was at a feverish pitch and on at least one occasion the railway companies made the mistake of continuing to work on the Sabbath, drawing the ire of Leitch who thought it improper to be depriving people of their religious holidays in favour of profiteering.

Perhaps presaging things to come, in 1844 Leitch took on some of the local "Patrons, Titulars and Tacksmen" regarding the "Teinds" of his parish. In plain English it appears that he was going to investigate the way that local funds, promised to the church, were being distributed.

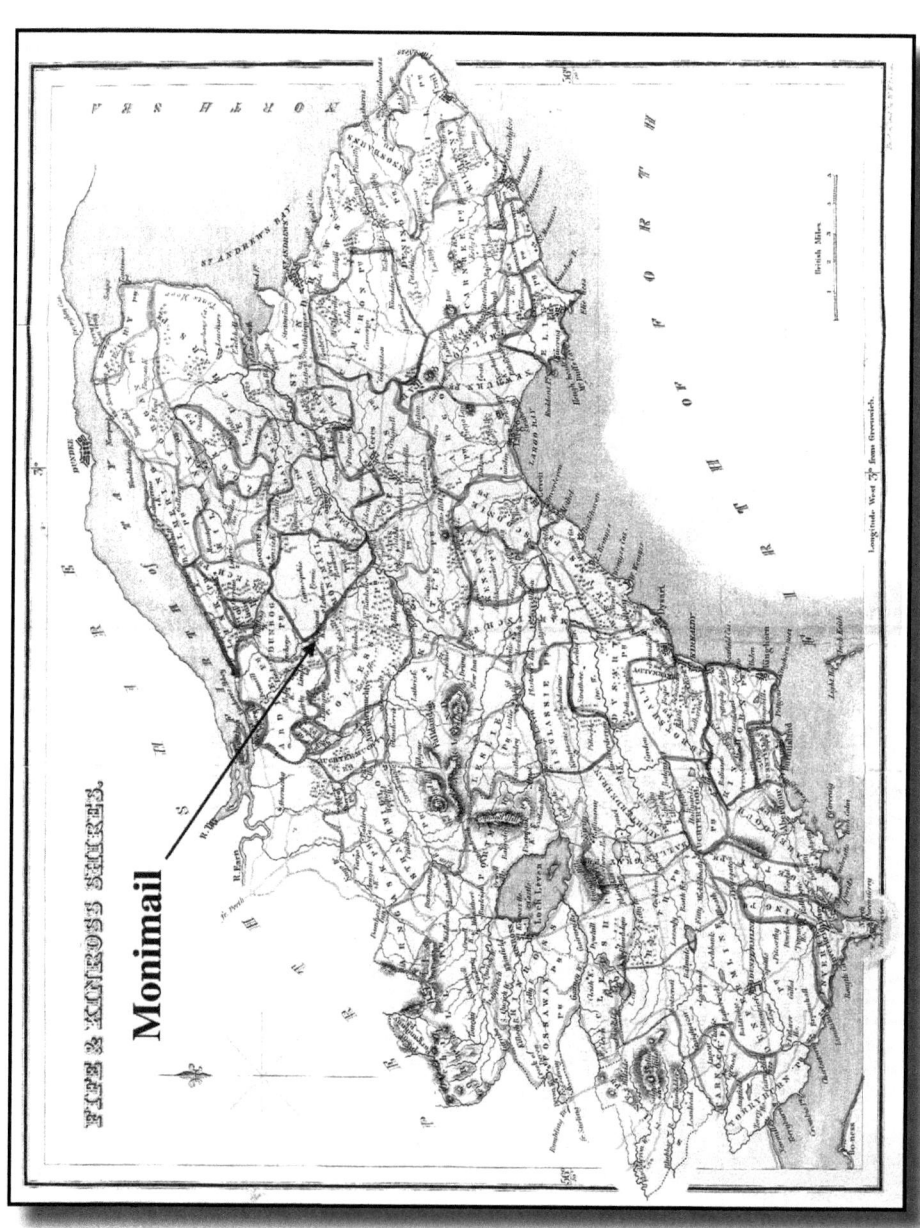

Map of Fife showing location of Monimail

In 1846 Norman Macleod presided at Leitch's wedding and for the next few years the couple lived at the Manse in Monimail where they started a family.[18]

As early as March 1850 Leitch was participating in a series of local science lectures in Cupar. His chosen subject at this time was astronomy. At the exact same time his old mentor John Nichol was also touring Cumbria with a series of well-publicized astronomy lectures. Leitch's first documented astronomy lecture was part of a series which took place at the Cupar County Hall. The speakers were an assortment of teachers and ministers, and the subjects included the excavations at Nineveh, social conditions in France, the crusades, and electro-magnetism. Leitch would speak on March 7th on the subject of the latest astronomical discoveries made by Lord Rosse's telescope. During this time Leitch was considered important enough that he was one of only a handful of private individuals in Britain who would receive a gratis subscription to the bulletins from the Royal Astronomical Society.[19]

In January of 1852 the local newspapers reported a "Testimonial to the Rev. Mr. Leitch". On this instance Leitch and his wife hosted their annual gathering in the local village hall in Letham. Apparently what was normally a friendly gathering of locals turned into a "large and influential" meeting attended by over 100 children and their families. Once again the people presented him with a gift for his services in teaching and for providing guidance for the children. This time the gift was a service of silver plate. Clearly Leitch was extremely popular with his flock.

By the end of 1852 Leitch was back again to participate in another series of lectures to be held at Cupar County Hall. The series would begin in December and continued until March of 1853. Leitch would be the second to last speaker. The subjects covered included Islam, fossils and paleontology, Scottish poets, crime and punishment, elocution, and the druids; while Leitch was to speak on electricity. To sit at the front would cost a local enthusiast one shilling to hear him speak, while a seat at the back was only sixpence. He spoke on the night of the 15th February and the reviewers commented, *"The Rev. Leitch delivered a most interesting lecture on Electricity, illustrated by various experiments. At the close he pointed out the bearing of recent discoveries on the progress of civilization, and the Christianisation of the world."*

Things were going well for Leitch, who frequently took on the role of moderator at the local Synod where his counsel seems to have been appreciated by his associates. Unfortunately his tranquil life in the quiet Fife countryside was about to be shattered.

William and Euphemia would have three sons and one daughter, but Euphemia would not live to see them grow-up. Two of their sons died while still infants, and then on the 8th of August 1853 Euphemia died at the age of 32, evidently from complications with childbirth. William was left as a single father to raise a four year-old and a five year-old.[20]

As was fairly typical of the times he elected to transfer custody of the children to his sister, who lived on the east coast at Saint Andrews. This would later prove to have a fortuitous side-effect and would elevate at least one of the Leitch clan to superstardom. But with the death of Euphemia, Leitch's enthusiasm for the public stage seems to have evaporated for about 18 months. He took up bee-keeping.[21] It was at this time he also entered the debate on the so-called "Plurality of Worlds" or in modern parlance, the possibility of life on other planets.

Notes:

1. The Presbyterian Historical Almanac, 1864

2. The Journal of Education for Lower Canada, Eighth Volume, 1864

3. The Canadian Naturalist and Geologist, Vol. 1, 1864

4. The Topographical picture of Glasgow in its ancient and modern state. R Chapman Glasgow 1822

5. The Canadian Naturalist and Geologist, Vol. 1, 1864

6. Paisley Herald May 28 1864, Fife Herald Jun 30 1864

7. Memoir of John Nichol, James Maclehose And Sons, Glasgow, 1896

8. Fife Herald Jun 30 1864

9. Ibid. Nov 7 1861

10. Ibid. May 26 1864, Dundee Advertiser May 27 1864

11. The Life of William Thomson, Baron Kelvin of Largs by Silvanus Phillips Thompson, MacMillan, London, 1910

12. Memoir of Norman Macleod by Donald Macleod, Worthington, N.Y., 1876

13. Ibid.

14. Men of Fife by M.F. Conolly, Inglis & Jack, Edinburgh, 1866

15. Fifeshire Journal, Nov. 11, 1861

16. Fife Herald My 26 1864

17. The Presbyterian Historical Almanac, 1864. John Pringle Nichol had also been a major public figure in Cupar in the 1820s.

18. Wigtownshire Free Press May 21 1846

19. Fifeshire Journal, Jan. 3, 1850

20. Ancestry.com

21. On the Development of Sex in Social Insects by William Leitch, British Association for the Advancement of Science Meeting September 1855, Murray, London, 1856

18th century philosopher Bernard le Bovier de Fontenelle, whose opinions on alien life affected William Leitch's work.

Chapter 3

Of The Plurality of Worlds - 1853

Ever since the invention of the telescope the subject of alien life on other planets had been a contentious one.

Newton's mathematical explanation of gravity published in 1686 all but shattered every previous cosmology. The inexorable truth so clearly laid out in his *Principia* made it hard to argue with his picture of the universe. One competitive system that had been in good standing for almost half a century was that proposed by the French philosopher René Descartes. Newton had studied Descartes at Cambridge. In his books *Discourse on Method* and *Principles of Philosophy*, Descartes had proposed a universe driven by vortices, in which each solar system was a vortex rubbing shoulders with the systems adjacent. Descartes held his own reasons for believing that everything in nature moved in a circular swirling fashion. His philosophy was complicated, but in his book *Le Monde* (The World), he advocated Copernicus' heliocentric solar system. He also tried to explain why the solar system had acquired its structure. In doing so he wrote down three laws of motion, which would have a profound impact on Newton's similar laws. However, after the persecution of Galileo and Bruno, Descartes was afraid to publish his "Cartesian" system during his lifetime, and so his book didn't see the light of day until fourteen years after his death. One of Descartes' most ardent disciples was the son of a Catholic lawyer from Normandy, his name was Bernard le Bovier de Fontenelle.

Born in 1657, Bernard was encouraged to follow in the family business and try his hand at legal work. After failing miserably as a lawyer he took up writing, which led him quickly to critical journalism. His wit and intellect rapidly earned him a place amongst high-society and his company was sought-after by people across France. His genius can scarcely be over-estimated for his time. He was a religiously enlightened catholic, standing up for the Protestants, who were about to be driven out of France in droves, and he understood that superstition and blind adherence to dogma was responsible for many of history's darkest moments. Bernard had a profound influence on Leitch.

The same year that Newton published the *Principia*, de Fontenelle published the first edition of a smash best seller called *Conversations on the Plurality of Worlds*. It is without a doubt one of the most astonishing pieces of scientific literature to appear in the 17th century. The book contains a series of fictional conversations between an unnamed central character (de Fontenelle incognito), and an inquiring and receptive Marquise who wishes to better understand the nature of the universe. Appearing, as it did, only a few decades after Galileo had been persecuted for his public musings on heliocentricism, the book could have been an enormous risk for de Fontenelle, especially since he chose to use Galileo's method to convey his message. He ingeniously disguised his theories behind layers of gentle persuasion and hypothetical suggestions, all in the form of a debate between the central characters. Bernard was however, not as vitriolic as Galileo had been in his book, instead of having his old-school disciple appear as a "Simplicio" (which had enraged Galileo's enemies) Bernard created a beautiful inquisitive noble-woman to play foil to his philosophical musings.

The book covers all of the controversial topics, including the possibility of alien life. The central character explained to the Marquise how the universe worked by frequently backing down on the more controversial concepts, gently leading her to the truth by constantly making allowances for old doctrines. The book is so cleverly written that it was all but impossible for de Fontenelle to be charged with heresy.

In his own words he said, *"What may be surprising to you is that religion simply has nothing to do with this system, in which I fill an infinity of worlds with inhabitants."*

One of the first steps he took was to explain to the Marquise the concept of relative motion; he did so by borrowing an analogy from Nicholas of Cusa.

"It's the same thing as if you went to sleep in a boat which was going down a river, you'd find yourself on waking in the same place and the same relationship to every part of the boat... I'd find the river bank changed upon waking, and this would make me see clearly that my boat had changed position... You know that beyond the circles of the planets are the fixed stars; there is our river bank."

As he explained the concept of the Earth rotating in a void he turned almost poetic, *"I sometimes imagine that I'm suspended in the air, motionless, while the Earth turns under me for twenty-four hours, and that I see passing under my gaze all the different faces... here cities with towers of porcelain, there great countries with nothing but huts; here vast seas, there frightful deserts; in all, the infinite variety that exists on the surface of the Earth."* (all translations here are from the wonderful 1990 edition by H. A. Hargreaves.)

He clearly understood the limits of the atmosphere, *"The air which surrounds the Earth only extends to a certain height, perhaps twenty leagues."* He also wasn't afraid to speculate on such things as the topography of the lunar far side, which he suspected would be much the same as the side facing the Earth. He quoted a lengthy extract from Ariosto's *Orlando Furioso*, which he used to lighten the moment, while deftly showing how outdated some of the old cosmology seems. He then turned to the nature of potential lunar inhabitants.

"I don't believe there are 'men' on the Moon. Look how much the face of nature changes between here and China; other features, other shapes, other customs, and nearly other principles of reasoning. Between here and the Moon the change must be even more considerable....He who would press on to the Moon assuredly would not find 'men' there."

He then said to the Marquise, *"I'll bet that I am going to make you admit, against all reason, that some day there might be communication between the Earth and the Moon....We're doing more than just guessing it's possible, we're beginning to fly a bit now (!); a number of different people have found the secret of strapping wings that hold them up in the air, and making them move, and crossing rivers or flying from one belfry to another...the art of flying has only just been born, it will be perfected, and some day, we'll go to the Moon."*

The previous extracts barely do justice to the sheer genius at work. De Fontenelle's masterpiece was an immediate smash hit. It was written for the general public in plain everyday French, even though he understood that this would probably alienate the philosophical elite. Such important matters were usually restricted to Latin. Now, for the first time in modern history, the average person could get a taste of cosmology and understand it.

He proceeded to throw out all of the old notions of lunar seas by suggesting that the famed dark patches are merely "great cavities." He dispensed with a lunar atmosphere by noting the absence of clouds. He predicted underground lunar cities by noting the hellish conditions on the surface and proposing that any indigenous life would almost certainly be driven underground. He speculated about life on the other planets, predicting that rather than being impossible due to the heat, any homegrown life on Mercury would surely freeze here on Earth. He even made allowances for life on the moons of Jupiter when he said, *"These planets are no less worthy of being inhabited for having the misfortune of being assigned to turn about another of greater importance."*

He also never shied away from extrapolating the big picture, *"Nature has held back nothing...she's made a profusion of riches altogether worthy of her. Nothing is so beautiful to visualize as this prodigious number of vortices, each with a sun at its center making planets rotate around it. The inhabitants of a planet in one of these infinite vortices see on all sides the lighted centers of the vortices surrounding them, but aren't able to see their planets which, having only a feeble light borrowed from their sun, don't send it beyond their own world."*

There are few accommodations in Bernard's book to appease the purveyors of scripture. Aristotle had always been the Greek most admired by the Christians of the Middle Ages. His universe was the version that fit snugly with Biblical doctrine. De Fontenelle made one gently sarcastic concession when he gave Aristotle an "out" for being so wrong, *"...it's impossible for Aristotle not to have held such a reasonable opinion (for how could a truth have escaped Aristotle?), but he never wanted to speak of it for fear of displeasing Alexander, who would have been in despair to see a world which he was unable to conquer. All the more reason for concealing the fixed stars and their vortices from him, if anyone had known about them in those times; it would have meant failure at court to mention them to him."*

The revelations of astonishing insight continue and one of them may have led (indirectly) to one of the great science fiction stories of our era; *Nightfall* by Isaac Asimov. In one of his many essays about himself, Asimov tells that the idea for his brilliant story came as a suggestion from his editor John W. Campbell. Campbell had been reading this quote from *Nature* by Ralph Waldo Emerson, *"If the stars should appear one night in a thousand years, how would men believe and adore, and preserve for many generations the remembrance of the city of God."*

Asimov's anecdote continues with Campbell suggesting that his young protégé write a story about a world that is exposed to true darkness only once in every two thousand years. Asimov took the idea and ran with it to produce one of the very best short stories in the history of science fiction. It appears that Emerson may have had de Fontenelle on *his* bookshelf. Speculating on a world located in the neighborhood of many stars Bernard wrote, *"So you see your sky glow with an infinite number of fires that are very close to one another and not far from you, and your night is no less bright than the day; at least the difference can't be perceptible, and to speak more precisely you never have night. They'd be quite astonished, the people of those systems, accustomed as they are to perpetual light, if they were told there were unfortunates who have true nights, who fall into very deep darkness and who, when they do enjoy the light, see only one solitary sun. They'd regard us as outcasts of nature and shudder with horror at our situation"* This short comment is almost the entire premise for Asimov's legendary story that won the author many accolades when it first appeared in 1941. It is all the more remarkable since the true shape and structure of galaxies was not confirmed for another two hundred years. Bernard's description must be an extremely close approximation of what you might see living on a world close to the center of a galaxy.

De Fontenelle is often mentioned in the early history of science fiction but, more often than not, it is in the wrong context. His great work was not really a work of fiction (even though the characters were imaginary) but a textbook for the masses. It was a brilliantly crafted explanation of the workings of the universe that stepped all over many sacrosanct religious concepts. De Fontenelle avoided the wrath of the church by couching the entire opus in the form of a hypothetical conversation. He never really stated, either directly or obliquely, that he believed any of it. He merely postulates. Many lesser writers would follow, borrowing his idea; and indeed many had preceded him with similar ideas, but de Fontenelle really brought the idea home to the public at large. His work is almost an antithesis of Newton because he didn't address the cosmos with the inexorable logic of pure mathematics, but from the viewpoint of plain old common sense. The fact that de Fontenelle still subscribed to the vortices of Descartes rather than the ordered mathematical certainties of Newton is irrelevant. It is his unadulterated passion for logic that really counted; placing him firmly in the ranks of 17th century genius.

Ever since De Fontenelle had published his most famous work, the discussion of alien life had seemed to be settled. Even Christian Huygens had published on the subject (posthumously and in favour). For well over 150 years most philosophers and scientists, at least in the world of Protestantism, had been content to accept the notion of other worlds with the potential for intelligent life.

However, in the last week of 1853 an anonymously scribed essay entitled *"Of The Plurality of Worlds"* appeared through the London publishing house of John W. Parker. It stirred up an immense controversy because it set about overthrowing everything that Fontenelle had said; using a deliberately provocative title. Over the next two years many scholars, theologians and astronomers once again took to the lecture circuit to state their own opinions on the controversial subject. William Leitch must have been one of the very first as he was apparently moved to speak up almost immediately. A series of ongoing lectures was already underway in Saint Andrews, which was about 15 miles east of Leitch's home in Monimail. On January 10th, which was, at the most, two weeks after the appearance of the anonymously authored essay, Leitch lectured on *"The Plurality of Worlds viewed in connection with recent astronomical discovery."* Having seen the night sky in all of its glory, Leitch was inclined to disagree with the mysterious author of the essay. He would go on to make this debate one of his favourite lecture subjects for the rest of his life. According to the newspapers his first talk was "handled well" and *"There was a respectable audience and all left, we are assured, highly satisfied with the truly popular manner in which the subjects were discussed."*

It is perhaps most important to take note of not just how early

Leitch took up the mantle of this particular fight, but also *where* he first spoke on the subject. His lecture was given at Saint Andrews. The entire population of Saint Andrews was only about 7000 people. One of its most distinguished residents in 1854 was Sir David Brewster, the famous scientist, who at that time was the Principal at Saint Andrews University. Leitch worked with both David Brewster and his brother George. He had served on the Evangelical Alliance Committee with David Brewster in 1846 and he had worked with George Brewster regularly at the Presbytery meetings for at least a decade. [1] They also shared the acquaintance of Norman Macleod who just a few months earlier had been at a party with David Brewster in Greenock. When Leitch came to Saint Andrews to speak out on the Plurality of Worlds it seems extremely likely that David Brewster would have been in attendance. It had been one of Brewster's favourite subjects since at least the 1830s.

Sir David Brewster (ca. 1850s)

Given the fact that the two men clearly knew each other well and may have previously spoken on the subject, one has to wonder how much influence Leitch may have had reigniting Brewster's passion. By the end of April Brewster was writing his own famous response to the anonymous essay, a book he was to call *"More Worlds Than One – The Creed of the Philosopher and the Hope of the Christian."* But seemingly unable to contain himself Brewster published an advance 26 page diatribe in the *North British Review* in May of 1854 in which he eviscerated the anonymous essayist. He gave no direct evidence that he knew exactly who he was attacking, but it soon became apparent that the author of the original anonymous essay was the well-respected scientist named William Whewell. Unlike Leitch's gentle approach, Brewster spared no effort to belittle the ideas which Whewell had put forth. Brewster had clearly read Huygens, adopting many of the same arguments about biology, but he also didn't stray very far from the assumption that the universe required a purpose. He insulted Whewell with "an ill-educated and ill-regulated mind" and for "panting for notoriety"; but despite making some valid points he was too strident in his ambition to refute Whewell. Amongst the many of Whewell's points he mistakenly disparaged were: the nebular hypothesis for the formation of the solar system; the idea of a goldilocks zone for life; the idea of the outer planets being super cold; the idea that planets around binaries may be in orbits non-conducive to life; the inherent variability of stars; and that the universe formed following basic laws and had been evolving ever since - without divine guidance. He even ridiculed Whewell for suggesting that the outer planets might be gaseous. One truly remarkable observation he made was the idea that a star's light might be obscured by the passage of comets, accounting for some stellar variables. As recently as 2015-16 anomalies detected by the Kepler space telescope found irregular patterns in the light from two distant stars. The three explanations put forth for this were (ironically) possible extraterrestrial life in the form of a Dyson Sphere, a proto-planetary nebula, or a swarm of comets obscuring the light. Whewell however preferred the simple explanation; that not all stars were like the sun. Brewster's book was published on June 3rd 1854 and by September it was into its third printing.

Whewell had been active in the cosmological debate since at least as early as 1833 when he released his book *Astronomy & Gener-*

al Physics Considered with Reference to Natural Theology. In this work he suggested that when Kepler first caught a glimpse of his great discovery he may have felt like he had been allowed to view part of God's great design. Kepler practically stated as much himself. As Whewell explained, *"When they had read a sentence of the table of the laws of the universe, they could not doubt whether it had had a legislator. When they had deciphered there a comprehensive and substantial truth, they could not believe that the letters had been thrown together by chance. They could not readily acknowledge that what their faculties had enabled them to read, must have been written by some higher and profounder mind."*

William Whewell (ca. 1860)

Whewell was a fellow of the Royal Society and Master of Trinity College in Cambridge. He may indeed have been perhaps the last of the classical philosophers. Ironically he was also the one who actually coined the term "scientist". He was the first to take modern empirical science and use it to paint a persuasive philosophical picture of man's moral superiority and uniqueness. He challenged the evolutionists, even before Charles Darwin entered the fray, with his 1845 book, *"Indications of the Creator"*. However, it was his essay *"Of The Plurality of Worlds"*, that represented his most persuasive thesis on man's distinctiveness in the universe. It is an extraordinary diatribe that uses all of the tools of modern science to show why he thought there are probably no intelligent life forms anywhere in the universe other than here on Earth. It was an early start to a campaign that was to be revisited by the physicist Enrico Fermi a century later when he famously asked the question, *"So where is everyone?"*

Whewell compared Geologic time with the size of the cosmos. In doing so he asserted that the Earth was a barren, void and lifeless planet for far more time than it had been populated, therefore the Universe at any given moment is probably most likely the same; due to its enormous scope. Although he clearly understood astronomical concepts he was a victim of the limitations of his era, in that he still confused nebulae with galaxies. His assessment, that a nebula was as tenuous as a comet, led him to the conclusion that all nebulae were uninhabitable. Unfortunately he could not grasp the enormity of the galactic structures in the universe. The distinction between nebulae and galaxies was yet to be clearly explained, although the German philosopher Immanuel Kant had speculated in 1755 that the Milky Way was a lenticular formation of stars which we were viewing on edge, and that Andromeda may be another such "Island Universe", as he called it. Thirty years later William Herschel had managed to confirm that the Magellanic Clouds were in fact vast clusters of stars and not gas clouds, but he couldn't confirm anything about more distant objects like the galaxy in Andromeda. To most astronomers of Whewell's time the glowing patches of light in the *deep* night sky were all the same; just tenuous gas clouds. Brewster correctly thought otherwise.

Whewell went on to assert that because true science deliberately shies away from speculation, a good scientist should be appalled by those who speculate, without proof, about life on other worlds. He then came to the conclusion that there was an ecosphere (a zone conducive to life) around the sun, with the Earth residing in the middle, and he concluded that Mars may be inside that zone and might be another abode for life in our solar system.

Whewell's use of language is gloriously persuasive and it is easy to see why he was so well respected in his time. For example he also expounded the theory that because man is able to comprehend such things as geometry and mathematics we must consequently be closer to the divine order of things because we are capable of understanding God's language.

"The Ideas according to which man builds up his knowledge, are emanations of the archetypal Ideas according to which the work of creation was planned and executed."

He addressed the basic requirements for life as we know it and suggested that those needs are not met by the other planets or by the sun and stars. He then looked at the concept of totally alien life:

"If gravity have not, upon any set of beings, the effect which it has upon us, such beings may live upon the surface of Saturn, though it be mere vapor: but then, on that supposition, they may equally well live in the vast space between Saturn and Jupiter, without needing any planet for their mansion. If we are ready to suppose that there are, in the solar system, conscious beings, not subject to the ordinary laws of life, we may go on to imagine creatures constituted of vaporous elements, floating in the fiery haze of a nebula, or close to the body of a sun; and cloudy forms which soar as vapors in the region of vapor. But such imaginations, besides being rather fitted for the employment of poets than of philosophers, will not, as we have said, find a population for the planets; since such forms may just as easily be conceived swimming round the sun in empty space, or darting from star to star, as confining themselves to the neighborhood of any of the solid globes which revolve about the central sun."

Whewell then concluded by turning the argument back against those who cited the enormity of the cosmos as proof that extraterrestrial life must be inevitable purely as a result of the sheer number of stars. He stated that the ludicrous waste of nature, in everything we see, such as how few seeds actually bloom, suggests the same potential for most of the stars to be similarly wasted.

The final conclusion is that Whewell left a raft of reasonable arguments that still spark debate today, as much in theological circles as in scientific circles. Unfortunately for Whewell his teleological view was being supplanted by a technological one. Almost immediately after publication his position was challenged by others in the world of academia, beginning with Leitch on a podium in St Andrews, and then by Brewster in the popular press.

Many of Whewell's arguments seem to have resonated with Leitch, but he disagreed with Whewell's conclusion. Rather than adopting the full frontal assault that Brewster had used, Leitch set about using multiple arguments in his lectures to refine his own stance. From the outset Leitch seems to have sided with the abundance of logic in favour of life on other worlds. Leitch clearly preferred Fontenelle's gentle persuasion to Brewster's stridency to explain this to people.

For the rest of his life Leitch found that his opinions on this subject were in demand; but it would be Brewster's more lavish rebuttal to Whewell which would steal the limelight.

In the spring of 1855, while still at Monimail, Leitch turned his attention to the world of Melittology; the science of bees. This must have fit in well with his understanding of botany and in that year he began to reach scientific conclusions about how he thought the humble worker bee might be able to change into a queen, thus preserving the longevity of the hive. His paper was published in the London scientific journals that year.

Around the same time he would take on another worthwhile public cause, the endemic lack of funding for the Royal Edinburgh Infirmary, a hospital designed to take care of ailing poor people. The hospital was short of funds and Leitch felt that it was the responsibility of all citizens to solve the problem, especially since the hospital was known to be training medical doctors who practiced all across Scotland. He must have taken the cause to his local benefactors because the entire Melville family and even the servants donated funds to the cause. Leitch also put his own money into the pot as the third largest single donor behind the Earl and the Countess. Today the Infirmary is still fully functioning as the oldest voluntary hospital and school of medicine in Scotland.

During that same spring Leitch was invited by the *Mechanics Institution* of Blairgowrie to deliver an address to the membership in the local chapel. The reception seems to have been favourable according to the Dundee newspapers:

"The Rev. Mr. Leitch, of Monimail, delivered a lecture in Brown Street Chapel on Wednesday evening to a numerous and respectable audience. The subject of this lecture was the "Plurality of Worlds", which the gentleman discussed in a most able manner. Several points

were illustrated by diagrams, which the lecturer provided for the occasion, and which made the lecture more than ordinarily interesting."

A more informative review appeared in a local paper:

"The ninth lecture in the series was delivered on Wednesday last, by the Rev. Mr. Leitch, Monimail, on the "Plurality of Worlds". The lecturer commenced by characterizing the present advanced state of astronomy, and casually noticing how early the idea was entertained of this world, so insignificant amid the immensity of creation, not being the only habitation of intelligent beings. The tremulous glimmerings of a previous age have now been amply confirmed by the startling rapidity of modern discoveries; and a plurality of worlds is not now the reverie of some solitary philosopher, but the firm belief and conviction of the majority of mankind. The lecturer then stated the principled arguments for a plurality of worlds: - the a priori *argument, the metaphysical argument, the analogical argument, and the scriptural argument. The two first favour the theory, the last countenances it, but it is principally founded: - annual and diurnal motions, moons, atmosphere, mountains, vales, light, heat, colour, and other physical appearances being common to all. After showing that the arguments lately propounded by Dr Whewell did not operate against the theory, he concluded with obviating any arguments which have been urged against the plurality of worlds, from the death of Christ in our own. The lecture was illustrated by diagrams, and was on the whole one of great excellence; and institutions may consider themselves very fortunate if they can secure the lecturer's able services. John Adamson, Esq., moved a vote of thanks to the lecturer in which he observed that the lecture delivered by Mr Leitch approached closely one delivered on the subject by Professor Nichol, and which he had the pleasure of hearing while sojourning in Glasgow."*

This latter review gives some insight into what had inspired Leitch to join in this particular debate. Not only does it tell us that he was working on his tutelage from Nichol but this review clearly describes how Leitch was aware of Whewell, who had sparked so much controversy and who was, to his dismay, now the most public voice in the discussion. Indeed, besides Brewster, Nichol was one of the few people still lecturing on the Plurality of Worlds during the "drought" years between Fontenelle and Whewell, suggesting his influence on Leitch's views on this subject probably predated Whewell's inflammatory missive.

Later in the summer of 1855 Leitch and several of his colleagues took a host of nearly 600 children on a field trip by railway to Crawford Priory, a crenellated mansion built for the Earl of Crawford, about 4 miles south east of Monimail. The challenge of learning how to deal with that many children must have come in useful when he became Convener of the Sabbath School Committee four years larer. The Sunday, or Sabbath, schools were an important part of the social life in Fife. Leitch worked hard for most of his adult life to bring a balanced education to people of all ages and stripes.

Crawford Priory (ca. 1880)

At the end of September 1855 Leitch was in Glasgow with his old friend Norman Macleod to attend the 25th annual meeting of the *British Association for the Advancement of Science*. This distinguished body had been created in the 1830s in response to a perceived elitism at *The Royal Society*. Somewhat ironically two of the prime movers in the formation of the group were the future adversaries David Brewster and William Whewell. On the morning of the 21st a group of about 400 scientists, ministers, gentry and philosophers, along with their wives, departed on a cruise aboard the steamer Iona from the Broomielaw docks to the Island of Arran. Their journey would take them through familiar waters for Leitch. Past Greenock where he had attended the grammar school before turning south and past his old home of Rothesay on the Isle of Bute. Aboard the ship were a distinguished cast of mid-century British scientists. At least three Fellows of the Royal Society, the Lord Provost, James Thomson, David Brewster and his brother George, Sir Roderick Murchison the famed geologist, at least one professor from the United States, one from Canada, the Principal of Glasgow University, and Leitch and Macleod.

The Broomielaw, Glasgow (ca. 1850)

The day seems to have been spent listening to lectures on the local geology as the ship passed each significant landmark before landing on Arran, whereupon the cast of enthusiasts went ashore armed with hammers for taking samples. After a meal at the castle of the Duke of Hamilton the steamer took them across the Firth of Clyde to Androssan where they caught the train back to Glasgow. A few years later their ship, the Iona, would sink just off Greenock.

Six months later Leitch was back on the lecture circuit, this time for the *Arbroath Scientific Association*. On the night of April 1st 1856 he earned this review:

"The lecturer before entering into the debatable question as to whether the planets were inhabited by sentient beings, gave an admirable description of the solar system. He referred to the different arguments that were used for and against the planets being inhabited, classifying them as follows: - the metaphysical argument, the analogical argument, and the scriptural argument. He seemed to think the analogical argument the most convincing of the three. The lecture was illustrated by several diagrams; and at the close a vote of thanks was cordially given to the reverend lecturer."

Clearly Leitch had barely changed his lecture since Blairgowrie, although the reviewer doesn't mention the *a priori* argument of a year earlier. This seems to make sense, because to say that such an argument, i.e. using reason or experience to apply to something which is clearly not proven almost seems redundant. His adoption of the analogical argument is by far the most illuminating in this instance because it shows that Leitch was clearly willing to take his stand based on logical extrapolation. This almost certainly was derived from his knowledge of the scientific method and his familiarity with mathematics and astronomy. Having impressed the academics of Arbroath, three weeks later Leitch took the case to some local students at Flisk, a small community on the River Tay about six miles north of Monimail, where he was to speak on astronomy to a Sabbath School gathering. A reviewer commented in the newspaper, *"The meeting was a very full one, and listened with breathless attention to the interesting and somewhat startling facts, which recent inquirers have brought to light in that noble science. The lecture was characterized by great clearness and simplicity of style, and evinced Mr Leitch's power of adapting himself to an audience, and carrying them along with him, even in the most difficult details. In the illustration of his subject he was greatly assisted by some beautiful diagrams of the planets and nebulae, which he exhibited. At the conclusion of the lecture, David Russell, Esq., Balhelvie, proposed a vote of thanks to Mr Leitch, it was seconded by the Rev. Mr Forbes, Moonzie, and cordially agreed to by the audience."*

Less than a month after his appearance in Flisk Leitch would launch himself fully into the world of politics. He wrote and arranged for publication a short essay under the unwieldy title of *"The Scottish Education Question: Being A Plea for the Denominational*

System, with Objections to the Parish Schools Bill of the Lord Advocate, in a Letter to the Right Honourable the Earl of Leven and Melville." It seems that this subject had been boiling around in political circles for at least six years when the Lord Privy Seal, Lord Melgund, a well-known Whig politician and previously the First Lord of the Admiralty, had introduced a bill into Parliament that would dismantle the schooling system in Scotland.

At the time that this bill was introduced most schools were run and maintained by the local parishes of the various fractured parts of the Church of Scotland. Funding was raised by the parishioners, the local ministers taught and examined the students (but not exclusively) and each parish relied on benefactors such as the local Earls and Lords to assist with funding the whole operation. The so-called Parish Schools Bill would dismantle all of that in favour of placing the entire system into the hands of a governing body, mainly centered in Edinburgh, but with oversight coming from as far afield as London. The bill was so controversial that it was attacked almost immediately by most of the different denominations, and also by most of the newspapers. The resistance was such that it would be in front of Parliament for years.

Needless to say this bill was an assault on everything that Leitch represented. He had come up through the existing school system, had excelled at Glasgow University and had gone on to his logical place in the world as a parish minister responsible for the education of his flock. It is fairly evident that Leitch was not typical of his kind. Many of his contemporaries didn't share his enlightened views about science or his strong belief that a man of God could live within a world where science confounded scripture on almost a daily basis. It is clear that Leitch was content, and even comfortable, within the system that existed and he saw no logical reason why the drive towards a secular school system would be an improvement. Despite his fascination with science he still had a streak of his grandfather's anti-burgher within him that railed against the state interfering with the church's role in society.

His May 1856 public missive to his local benefactor, Earl Leven, would be his most visible volley in defense of a fully rounded education which included theology. He was now very much in the fight and this whole issue would follow him for the rest of his life, even to Canada. Almost immediately the partisan *Fife Herald*, known for its long-standing support of the bill, fired back at Leitch. However, it seems that the worst accusation that they could muster against him was that he said "live and let live," i.e. create your secular schools as you wish, but leave us alone.

The politics of this argument were certainly more complex than this brief description; the endowments already in place from wealthy benefactors were dependant on the existing system; but Leitch, a minister, could hardly have been expected to take a more reasonable position.

Despite this important distraction, Leitch continued to live his life without obvious contradiction. He preached to his community, he taught the local children, he lectured at home and abroad on scientific matters and seems to have been comfortable wearing the mantle of scientist/priest.

On Thursday 25th November 1856 at the invitation of the dauntingly titled *Committee of the Total Abstinence Society of Paisley* he held forth with his latest astronomy lecture near to his old haunts in Glasgow.

"On the evening of Tuesday last, the Rev. Mr Wm. Leitch, of Monimail delivered the first of two lectures on Astronomy, and handled his subject in an entertaining and discursive manner, interspersed with anecdotes and reminiscences. He explained the method of ascertaining the density or weight of the various planets, illustrated their comparative sizes and distances, showed the construction and powers of the telescope, and finally and preparatory to his next lecture on the Plurality of Worlds, adverted to the spots on the sun, and the nature and appearance of the moon, illustrating the latter by an illuminated diagram. A deal of interest attaches to his next lecture, which takes place on Tuesday first; and from the ability of the lecturer, and the fascinating nature of his subject, a treat of no ordinary kind may be expected."

The following week he returned to a similarly warm reception:

"The Rev. Mr Leitch delivered his second lecture on astronomy on Tuesday evening last, to an interested and approving audience. He enumerated the several arguments brought forward in support of

the theory of the plurality of worlds, adverted to the various persons who had given their support to the rival systems, and adduced the illustrations on which they ought to rest their statements. Many curious and interesting facts were thus brought before the audience, and a display of illuminated diagrams enhanced the interest of the remarks. At the close a hearty vote of thanks was awarded Mr Leitch."

In his role as wandering lecturer, between 1840 and 1860 Leitch wrote and spoke on a diverse range of topics, including artillery and other projectiles.[2] His lectures on different aspects of natural philosophy would be informative and would serve to elevate his reputation in the world of science. He seems to have been quite competent and his lectures were clearly well organized. The reference in these two reviews to "illuminated" diagrams is probably explained by the use of the so-called "magic lantern" a kind of early slide projector. John Emslie, a London artist had been producing popular "see-through" slides for this exact task since 1850. Leitch had been on hand for a science lecture given to some students at the Sabbath school in Letham where a magic lantern was used to show the children diagrams of the earth, sun, moon and planets. That same lecture also demonstrated an air pump and the effects of magnetism. Leitch thanked the visiting speaker and stated that *"some might entertain the opinion that such exhibitions were apt to lead the young mind away from the more appropriate instruction of the classroom, but he hoped the audience and public would understand in this instance that such was not the fact, as these exhibitions were only meant to be subsidiary to the usual and more common branches of education."* Clearly he was not only content for science to be taught to a denominational Sabbath school, he was advocating for it.

By February of 1857 Leitch's talk on alien life and astronomy elevated him to the lecture series planned by the prestigious Watt Institution. Located in Edinburgh the Watt Institution had been established as a counterpart to the Anderson Institution of Glasgow. The Andersonian claims to be the world's first technical college, while the Watt Institution claims to be the world's first mechanics institute. Leitch had lectured at the Andersonian on mathematics while attending there in the 1830s. Both schools still flourish today, albeit with slightly different names. This particular lecture series took Leitch to Dundee.

Once again the newspapers reported on his performance, this time clearly stating his own opinion on the debate.

"The Rev. Mr Leitch of Monimail delivered the second lecture of the course on Wednesday evening, his subject being "Recent Astronomical Discoveries, viewed in reference to a Plurality of Worlds." The lecture was exceedingly interesting, the conclusion arrived at being that the probabilities were on the whole in favour of the planets being inhabited by intelligent beings."

Despite these public discursions the accusation of being hidebound or dogmatic would stalk him for the rest of his life. The notion of an open-minded Presbyterian was simply too strange for his political and theological opponents.

The Parish School Question - 1857

One subject which had been close to his heart was the missionary work being done by his church in India. Leitch had been involved with this since at least 1850, so when the overhaul of the parish school system became imminent he began to look at how such an overhaul would affect education at the missions in India. He wrote a 70 page essay on the subject which was published in March 1857 by his old friend Myles MacPhail of Edinburgh.

The British East India Company, which was essentially governing India, was willing to extend monetary grants to the Church to assist in the establishment of schools to teach the various Christian doctrines, but they wouldn't do it to the detriment of secular schools. Some people in the Church of Scotland argued that they should not take the grants as this was tantamount to endorsing secularism. Leitch thought otherwise. It would be in this important argument that he would reveal his ability to perceive a third path, something that he would hone and sharpen and apply in his public discourse on science versus religion.

In this case the choice was being presented by his opponents as, *"If we accept the Grants we are sanctioning the secular schools in India."* He argued that this was an erroneous conclusion. That one did not, by default, lead to the other. He argued that it was not the church's duty to railroad the state into depriving other denominations, or atheists, of support.

By assuming this more subtle position he was not considered to be fighting the righteous evangelical fight that some of his church brethren demanded. However, it seems that Leitch was not lacking in conviction. By taking the church out of a rounded education, as had been done in the USA, he felt that they were breeding a generation of "infidels." He also argued that the solution was not to allow the state to take over control of religious education, citing the examples of Canada, Prussia and elsewhere as examples of how that can go wrong. In the case of Prussia the teachers were often agnostic or atheist and were doing more harm than good to the church. His argument was that if the state, or alternatively some forced conglomeration of opposing religious bodies, was left to provide religious instruction it would fail miserably. In his mind it was right and proper that the church be in charge of religious teaching and that, rather than forcing the state to side solely with the Presbyterians or any other denomination, it was more logical to take the money, allow others to take their share of the money, and go about their mission.[3]

This argument would follow him to Canada but in this document Leitch's opinions are clearly compatible with those taken by his church in 1849, that, *"Instead of arrogating an exclusive right to control the religious education of the whole people of Scotland, she (the Church of Scotland) proclaimed her willingness that other Churches should receive State aid, and co-operate with her in the great work of the godly upbringing of the young."*

Although Leitch made it clear that he disapproved of Catholicism he also understood that the government couldn't be seen to be taking sides when it came to missionary work in India. He also made it clear that trying to ram Christianity down the throats of the Indian citizenry would do more harm than good and that it was better to just create Christian schools alongside secular and Hindu schools and allow people to come to them, if they so desired. Meanwhile, the government would be free to inspect the Christian schools to make sure that the needs of the state were still being met, i.e. teaching mathematics, language etc., but that:

"The government at once conceded this as reasonable, and gave strict injunctions that, even though the religious teaching should be carried on in his presence, the government inspector was not to presume to report upon it."

In this he again proved that he was consciously advocating for a rounded education.

"The government of India are, like a parochial board, a purely secular body, and they are anxious to bestow temporal gifts upon their poor subjects. They, for example, wish to bestow a knowledge of agriculture, the want of which is so frequently the cause of famines: must they say to the natives, 'Unless you take our religion, you will get no agricultural schools, and so be left to perish from your ignorance?' Anything but religious neutrality in such a case would be a grievous wrong.

"Christianity is regarded as the divinely appointed element for developing the moral, religious, and intellectual nature of man; and the mechanical and secular branches of education are only the instruments through which the great end is attained.

"The subtle intellect of the Hindoo; his deep reverence for learning; the high standard of education which many of the natives now receive; the fact that their religion is very much a matter of literature and science;—these, and other circumstances, all combine in demanding a learned ministry for India."

Perhaps the single phrase which most sums up his attitude to life would be this one:

"Secular knowledge is God's truth as well as the Bible; and it would be the greatest libel on the latter to hold that there is a necessary conflict between them."

Although this lengthy defense of the Scottish model of parish schools was published in Edinburgh, there can be little doubt that the substance of it would be quietly filed away by his evangelical opponents who would later continue to doubt his motives when he came to the defense of the denominational school system in Canada.

Two weeks later his name was back in the papers again, but this time on the subject of sun spots. *The Paisley Advertiser* ran a comment about his recent astronomy lectures where he had apparently discussed a popular theory that the price of corn seemed to fluctuate in accord with sun spot cycles. The article doesn't say that Leitch specifically believed this idea, but it very carefully compared the price of corn relative to the 11 year solar cycle and said that there might be some truth to the notion. From this we can see that Leitch was delving into all aspects of astronomy, including the very real effects of other celestial bodies upon life on Earth.

Through the summer of 1857 Leitch continued to play his role within the small community of Monimail. Working with the Countess of Leven he would continue to take the school children on field trips to Melville house where they would be run through their tests and returned home well fed. This seems to have been typical of the role for the Presbyterian minister which he was now defending so publicly. He also would deliver the eulogies at the end of each year for the people who had died in his Parish.

In March 1858 he was back on the lecture circuit. On the 24th he spoke at Falkland. The reviewer clearly felt that Leitch was presenting an open-minded position when it came to the subject of alien life.

"The lecturer commenced by relating the notions of the ancients as regards the heavenly bodies, particularly the earth; he then showed how the distances and weights of the planetary bodies were ascertained. He spoke of the great discoveries made by aid of the telescope, and showed by a number of beautiful diagrams the various appearances of the planetary bodies when viewed through that instrument. The lecturer treated the arguments pro and con with considerable ability, and he took great pains to render his observations as lucid and as simple as possible, so that the most ignorant could not fail to comprehend them."

Another issue which Leitch seems to have felt strongly about was the education of women. On June 1st 1858 he attended the *20th Annual Meeting of the Scottish Ladies Association for the Advancement of Female Education in India*. This meeting seems to speak for itself and is entirely consistent with his essay of 1857.

In September of 1858 Leitch took a three week trip around Europe at the invitation of Norman Macleod. They visited many of Macleod's previous haunts in Germany, France, Switzerland and Holland. It was a melancholy nostalgic excursion for Macleod who reported that Leitch was always there to bolster his spirits. [4]

While Leitch had earned a place of respect in both the church and scientific communities his old college friend Norman Macleod had risen to the highest ranks of his profession, ministering to the Queen.

In 1859 an ambitious young man named Alexander Strahan, living in Edinburgh, had taken notice of Macleod's preeminence in the community. Strahan wanted to enter into the world of magazine publishing, and so he approached Macleod to be editor for a new Christian magazine. It was to be called *Christian Guest: A Family Magazine for Leisure Hours and Sundays*. What might seem odd to modern sensibilities is that in 1859 it was considered highly inappropriate to read anything other than the Bible on Sundays. Strahan was advocating a somewhat minor revolution by suggesting a Sunday magazine, but he had heard that Macleod was particularly open-minded. For reasons we will likely never know Macleod agreed to work for Strahan for nothing more than a share in whatever profit Strahan might squeeze from the whole operation. The first issue was released in February 1859 and it ran weekly until the end of 1859 (a total of 46 issues) before it became evident that a more ambitious publication was possible. In the last issue Macleod and Strahan announced their intention to "fold" *Christian Guest* into a more expansive magazine to be called *Good Words*.

Only one small article had appeared in *Christian Guest* that might have foretold what was to come. It was called "*The Bible on Physical Science*". The article is unattributed and may well have been written by Macleod, Strahan or even Leitch. In only a few hundred words it opened the door to the notion that discussing science was not heretical, since the Bible said virtually nothing on the subject of the physical sciences, despite mentioning virtually every aspect of the physical world: birds, fish, animals, stars, planets, etc.

The *Christian Guest's* most significant major side-benefit was that Macleod developed a fondness for Strahan and was willing to continue their relationship. In an advert at the end of 1859 they unveiled their stated goal of providing a forum for civil and intellectual discourse from all denominations and by *"many of the best known writers of the day."*

Macleod had risen rapidly in the ranks of his peers and was now making a name for himself, not only in the burgeoning world of magazine publishing, but he had befriended the most powerful monarch in the world. Queen Victoria was in her 40s and had been sitting on the throne of England, Scotland and the rest of the Empire since 1837. Unknown to Macleod just as he was reaching a level of celebrity considered by many to be inappropriate for a man of the cloth, the Queen would lose her consort Prince Albert and spiral into a depression which she remained in for much of the rest of her life. Very few had access to her, but Macleod was one of the privileged inner circle.

By this time Leitch had been ministering to his parish in Fife for 16 years and was pursuing a series of scientific interests. He had given lectures on astronomy, he still kept his own beehives as a science experiment and had published at least one paper on that subject; and he had kept in touch with the scientists at Glasgow University, including William Thomson. He also took an interest in ballistics and particularly the new science of spin-stabilized projectiles.[5]

The Minié rifle was the latest development in weaponry. Instead of using the traditional round musket ball it used a soft-skirted slug with one end streamlined and the other with a concave opening. When combined with the rifling of the barrel the expansion of the gases made the skirt expand and as the slug left the barrel it was set spinning. This increased the range and accuracy and made the Minié rifle an extremely efficient killing machine. Interestingly the same spin-stabilization had been introduced to the war rocket by a British inventor named William Hale just a few years earlier. Hale was still introducing improvements to his patents as late as the 1860s but essentially his gunpowder rockets expelled the hot gas through a series of vents which caused the rocket to spin and thus fly true. The ability to do away with the long stick, which had traditionally been employed to make rockets fly straight, was a significant technological advance. Understanding these two devices certainly shows that Leitch grasped the Newtonian principles at work. Perhaps his work on steam propulsion at Glasgow may have seeded this interest. He seems to have had a more than average understanding of engineering since he later was credited with inventing a new kind of device which was used to heat some of the churches in Monimail and elsewhere. It is not yet known exactly where or when Leitch lectured on artillery or the Minié rifle but there are at least two references in the records of the time where his colleagues praised his lectures on the subject.[6]

It is also not known if Leitch attended the first Great Exhibition in London in 1851. It seems likely, since there was a copy of the event program in his library and it was the most ambitious fair of the era, attracting untold thousands of visitors. One of the most highly visible displays in the main exhibition area was a giant 11" refracting telescope on a massive stand. It stood fully 25 feet high and was ultimately destined to be installed in an observatory in Scotland. The subsequent 1862 exhibition was even larger and both the Minié rifle and the Hale spin-stabilized war rocket were exhibited almost side by side. Again we don't know if Leitch attended in the summer of 1862 but it does serve to illustrate that these weapons were highly visible new additions to the arsenals of the world. Like all of the local ministers and priests it seems inevitable that Leitch would have been involved in the move to raise a volunteer army in Fife in 1859. During that summer troops in kilts carrying bagpipes could be seen proudly marching all across the country lanes of the Scottish counties.

Ships in the Night - 1859

Just 30 miles west of Monimail was Inzievar House, the home of Archibald Smith Sligo. Inzievar still exists today as a well-groomed gothic manor house converted into apartments, but in 1859 Archibald Smith inherited the house when he married the daughter of George Sligo and took on the name of his wife's family. Smith's brother was the Reverend William Smith of St Mary's Catholic Church in Edinburgh. Their father was the Secretary of the Catholic Institute of Great Britain. Clearly they were very highly placed in what remained of the Scottish Catholic hierarchy. Leitch would certainly have known the Smiths by reputation. Both were involved in the raising of the volunteer army.

Since his wife had died Leitch's surviving two children still lived in the town of Saint Andrews. In the years ahead his family would become much more entrenched in that coastal town. Clearly Leitch was well-known there, not least from his lectures but he had also delivered at least one sermon there in 1846.[7]. The Reverend William Smith would go on to become Archbishop of Edinburgh and Saint Andrews, a title which had its ancestral home in none other than the tiny hamlet of Monimail. It is likely that most of the Presbyterian ministers were familiar with their suppressed neighbouring Catholic counterparts but in this instance it seems extremely unfortunate that the Presbyterians didn't fraternize with them. Regardless, in part due to a widespread distrust of Catholicism, a great opportunity eluded history in the summer of 1859.

On August 26th of that year a relatively unknown French playwright by the name of Jules Verne arrived by ship at Liverpool docks. On the 27th he took a train to Edinburgh and arrived late at night in the pouring rain. He checked into Lambre's Hotel on Princes Street which prided itself that its employees "Parle Francais". The hotel was located directly across the street from the Edinburgh General Railway terminus. On the 28th Verne and his travelling companion toured Edinburgh and made a trip to Portobello beach. About half a mile from Verne's hotel was Smith's church.

(Verne would later describe his trip to Scotland in a fictional account which was rejected by his publisher. Had it been accepted, it would have been his first novel. It would finally be published in 1987.)

Smith's Forth Iron Works (ca 1860)

On the 28th the Reverend Smith met the aspiring young French novelist who was in Scotland on his first trip to the ancient ancestral home of his mother's family. Smith persuaded Verne that he should come and visit him in Fife at Inzievar House in Oakley (which Verne called Oakley Castle) and Verne accordingly obliged him. Verne's visit to Edinburgh happened to coincide exactly with the moment that two emissaries arrived from Queen's University in Canada to meet with William Leitch.

By this time Leitch had published enough papers and articles to be elevated to the Committee of the Church Union in Scotland. He would continue to rise in the ranks of the Church until the arrival from Canada of the Reverend Thomas Barclay accompanied by Alexander Morris, a well connected member of the Kingston Ontario community.[8]

On the 29th of August Verne described taking a steamer up the Firth of Forth from Edinburgh in very bad weather and was wondering if he hadn't made a terrible mistake. Eventually he and his companion were rowed ashore where they arrived "sick and wet". They were greeted by Reverend Smith and several friends who walked them across country to the village of Oakley where the Smith family ran a huge Iron Foundry. Archibald Smith had at that very moment gone off to Edinburgh to assume his role in the national mobilization of volunteers to create artillery and rifle divisions. In early 1860 he became Captain of the Edinburgh Volunteer Rifle Corps and he held periodic drills and inspections at Oakley.

One can readily assume that Smith's foundry would have been turning out armaments of all kinds for the newly minted troops including the revolutionary Minié rifle. Verne marveled at the huge industrial works and stayed one night at Inzievar, which it was later said was the inspiration for the inside of Captain Nemo's submarine.

Remarkably we know that William Leitch was definitely in Edinburgh on Tuesday the 30th August and very probably traveled there the day before to be in time for his meeting with Morris. The Presbytery of the Church of Scotland usually had its meetings at the Synod Hall on Queen Street which was literally 350 yards from Verne's hotel and almost within sight of St Mary's. Leitch met with Morris in Edinburgh on the 30th.

At that time Verne was not yet a household name and he hadn't begun his *Extraordinary Adventure* series. In 1859 it was Leitch who had the subject of outer space foremost in his mind, a subject which had not yet crossed the mind of the man who would become the most famous inspiration in the history of spaceflight.

It is fairly certain that Leitch and Verne didn't meet, but the incredible fact is that they were very probably within a few yards of each other on that rainy day of August 29th 1859, possibly even passing in the railway station. Since it was Verne's only trip to Scotland during Leitch's lifetime and Leitch lived over forty miles away on the other side of the Firth of Forth and apparently didn't make frequent trips to Edinburgh it is astonishing that they came within such a short distance of one another, *and that we can prove this*. On the 30th Verne travelled from Fife to Glasgow by train, Morris made the same journey six days later after his meeting with Leitch.

An Invitation to Canada - 1859

The Reverend Barclay and his companion Alexander Morris had travelled to Scotland from Toronto looking for a replacement to fill the position of Principal at Queens University. Alexander Morris was highly placed in the confidence of John A. Macdonald, destined to become the first Prime Minister of Canada. Morris would later be instrumental in the Confederation of Canada when he managed to broker a peace between the leaders of the two political parties, Macdonald and George Brown, but his goal at this time was to fill a post at Queens which had remained vacant for 14 years. Not long after he arrived Morris was directed towards Leitch by an assortment of respectable members of the Church of Scotland, not least, Norman Macleod.

On August 30th 1859 he had his meeting with Leitch in Edinburgh. The letter in Morris' papers at the Toronto University Library indicates that Leitch was hesitant to come to Canada for *"the teaching of dogmatic theology,"* but evidently Morris was persuasive.

Norman Macleod officially nominated Leitch for the prestigious position, due to his *"high theological and scientific attainments, active missionary spirit, earnest Christian character, and urbanity of temper."* [9] The news at the time went on to state that he was *"moreover, distinguished for his proficiency in astronomy and the natural sciences generally, having taught them in the University of Glasgow."* [10]

His friends encouraged him to take the position in Canada because they noticed how sad he had been since losing his wife. Less than three weeks after Morris spoke to him Leitch also lost his mentor, John Pringle Nichol, who died on September 19th at a convalescent home just a few hundred yards away from the Leitch family residence on the Isle of Bute. Nichol had been Leitch's mentor and friend and, combined with the loss of two children and his wife, this new bereavement may have been the trigger that caused Leitch to succumb to Morris' solicitations.

A letter from Morris to Leitch (30 Aug. 1859)

Morris and Barclay would soon return to Canada and at a meeting of the Board of Trustees of Queen's College they were congratulated for having recruited such a well-respected man of science who could also be trusted to represent the church. In November the Board approved the choice of Leitch and all that was left to do was for the man himself to show up. It would take another year before he would report for duty. But that year would not be wasted.

During the first week of December 1859 the newspapers across Scotland reported the news that Leitch had been appointed to Queen's University in Canada. The Greenock Advertiser also noted in the same story that *"Good Words - A weekly journal of instructive reading is to be commenced in the 1ˢᵗ January, under the editorship of Dr Norman Macleod of the Barony. The magazine is to be religious in its tone and tendency, but will have no denominational association."*

Leitch's articles had already appeared in *MacPhail's Edinburgh Ecclesiastical Journal,* Kitto's *Journal of Sacred Literature, The Edinburgh Christian Magazine, The Scottish Review,* and most importantly in 1859 he had now been invited to write for the new magazine being edited by his old school friend Norman Macleod. [11]

While pondering his decision about moving to Canada Leitch began the composition of a series of articles on astronomy for Macleod's upcoming magazine. This string of essays would be spread out over the period stretching from January of 1860 until December of 1862. Ultimately, at least one of these essays should earn Leitch a distinguished place in the history of space travel.

Notes:

1. Evangelical Alliance, Northern Warder and General Advertiser, August 13 1846

2. Men of Fife by M.F. Conolly, Inglis & Jack, Edinburgh, 1866

3. Leitch later developed a better understanding of what was happening in Canada. After he moved there he became friends with Dr Ryerson who was in charge of the Canadian educational system. Ryerson had left the schools under secular control because the churches simply didn't have the resources to cope with running them. However, once the Catholic church asked for grants to run their own schools it became inevitable that the Protestant denominations would soon follow suit. See: A Winter in Canada by William Leitch, Good Words, Strahan, London, Dec. 1862

4. Memoir of Norman Macleod by Donald Macleod, Worthington, N.Y., 1876

5. Men of Fife, M.F. Conolly, Inglis & Jack, Edinburgh, 1866

6. Ibid.

7. Swallow Street chapel on August 23 1846. Evangelical Alliance. Report of the proceedings of the Conference - 1847

8. Globe, Toronto, Dec. 13 1859

9. The Bulletin Records and Proceedings of the Committee on Archives of the United Church of Canada, No. 9, United Church Publishing, 1956

10. MacPhail's Edinburgh Ecclesiastical Review, 1859

11. Memoir of Norman Macleod by Donald Macleod, Worthington, N.Y., 1876

Leitch & Macleod September 1858 [4]

Chapter 4

Good Words - 1860

When the first issue of *Good Words* hit the streets on January 1st 1860 it was to appear as a weekly (just like *The Christian Guest*) but it was selling for four times the price. It was contemporaneous with such other British magazines as *"The Cornhill"* and *"Chambers Journal"* and *"The Atlantic Monthly"* in the United States. Indeed the first issue of "Good Words" appeared on exactly the same day as the first issue of "The Cornhill".[1]

The high price was a gamble taken by Strahan because he knew that this might put it out of the reach of the general public. However, Strahan would take similar gambles for the rest of his professional career. In the case of Good Words, it paid off...or at least it paid off for the first decade.

Macleod had been fond of William Leitch since their school years together. He had presided at Leitch's wedding and he must have been keenly aware of Leitch's anti-burgher upbringing as well as his remarkable propensity for hard work. Leitch had sat for many a freezing night making careful measurements in Nichol's observatory in Glasgow. His understanding of astronomy was on a par with any of his peers and his tendency to try and explain the physical universe without conflicting with scripture fit in perfectly with Strahan's attempt to reach a wider audience with Good Words.

Strahan knew that he was running the risk of alienating the church by publishing a magazine that he was actively soliciting as predominantly theological and pious, whilst still including content that was clearly of a more secular nature. He also knew that the schism which had developed between the evangelical wing of the Church of Scotland and the more liberal members (including Macleod and Leitch) had the potential to be far more damaging to his aspirations. Macleod's name on the masthead was a double-edged sword. Strahan had to deliver carefully selected content in the new magazine so that the sword cut in his favour.

In the very first issue there was an unattributed column called *"Sketches in Natural History"* which seems clearly to have been written by someone with knowledge of astronomy, zoology, botany, geology and biology. This article appeared on page 10 of the very first issue of Good Words and set the stage for discussions of what might be perceived as radical scientific theories, inconsistent with a typical theological magazine. In two pages this article cited Chalmers Astronomical Discourses, the existence of billions of foraminifera making up the bulk of the world's limestone, the studies of Ehrenberg into the nature of rock fossils, the observations of Erasmus Darwin (his book "Love of the Plants") and James Hooker and the respiration of plants. The subject matter and language seems to fit Leitch's interests almost too perfectly. Since he often wrote anonymously it seems highly probable that it was written by him. It may just as easily have been written by Macleod or Strahan or any number of like-minded individuals, but it establishes that Macleod and Strahan were going to take science seriously in their new magazine.

Articles in *"Good Words"* selected by Macleod ran the gamut from the caravanserais of Iraq to anecdotes about the Emperor Joseph II. In its first year *Good Words* was a weekly magazine and Macleod would serialize Leitch's essays.

In the second week of January 1860 the first article identified as being written by Leitch appeared under the banner heading *"God's Glory in the Heavens - The Teachings of the Stars. No 1. The Moon - Is it Inhabited?"* There can be little doubt from this title, and the subsequent article, that Leitch was about to embark on a science lesson about alien life, and because of his credentials as a respected minister, Strahan could at least expect that the clergy might give him the benefit of the doubt. Leitch was a safe bet. Both Macleod and Strahan could rely with some degree of certitude that Leitch was not about to fly off on any sacrilegious tangents.

Beginning with that second issue, and then approximately once a month in 1860 until Leitch's departure for Canada in October. The subjects were as follows:

- The Moon – Is it Inhabited? (#2, Week 2, Jan 1860)
- The Moon's Invisible Side (#6, Week 4, Feb 1860)
- Lunar Landscape (#11, Week 3, Mar 1860)
- Discovery of the New Planet Vulcan (#15, Week 3, Apr 1860)

- The Approaching Total Eclipse of the Sun (#19, Week 2, May 1860)
- Comets – Their Nature and Design (#30, Week 5, July 1860)
- Comets – Their History (#33, Week 3, Aug 1860)
- The Sun – Its Work and Structure (#37, Week 3, Sept 1860)
- The Structure of the Planets (#40, Week 2, Oct 1860)
- The Nebulae (#46, Week 4, Nov 1860)

Just prior to Leitch's departure for Canada the publisher of *Good Words*, Strahan (based in Edinburgh and later, London), evidently negotiated a deal with him to anthologize these essays about astronomy into a book, which was to be entitled *"God's Glory in the Heavens; or Something of the Wonders of Astronomy."* Advertisements appeared announcing the forthcoming edition at the end of October 1860.[2]

Leitch's attempts to reconcile scientific observation with religious doctrine were quite atypical for a minister of that time. Such attempts were fraught with risk since the Church of Scotland had been fractured for decades and was undergoing yet another schism between the evangelical factions and those who were less conservative. Macleod was firmly placed in the latter group and, even just by association, Leitch would go with him. Over the next two years this would put Leitch into the awkward spot of not being pious enough for some in Scotland and too pious for some in Canada. Alexander Strahan had tried to position Good Words as a forum for all voices to be heard and Leitch's predisposition for science, while remaining a reliable minister, made him a perfect contributor. His most industrious attempt to explain the apparent contradictions between science and scripture seems to have manifested itself in the series of essays written for Strahan and Macleod.

It is important to understand the world in which Leitch was writing. Several references from the period suggest that Leitch and his brother were acquainted with the work of some of the most respected scientists of their time. If the younger Leitch didn't know these men first-hand he was certainly keeping his pulse on the scientific breakthroughs coming from London and elsewhere.[3]

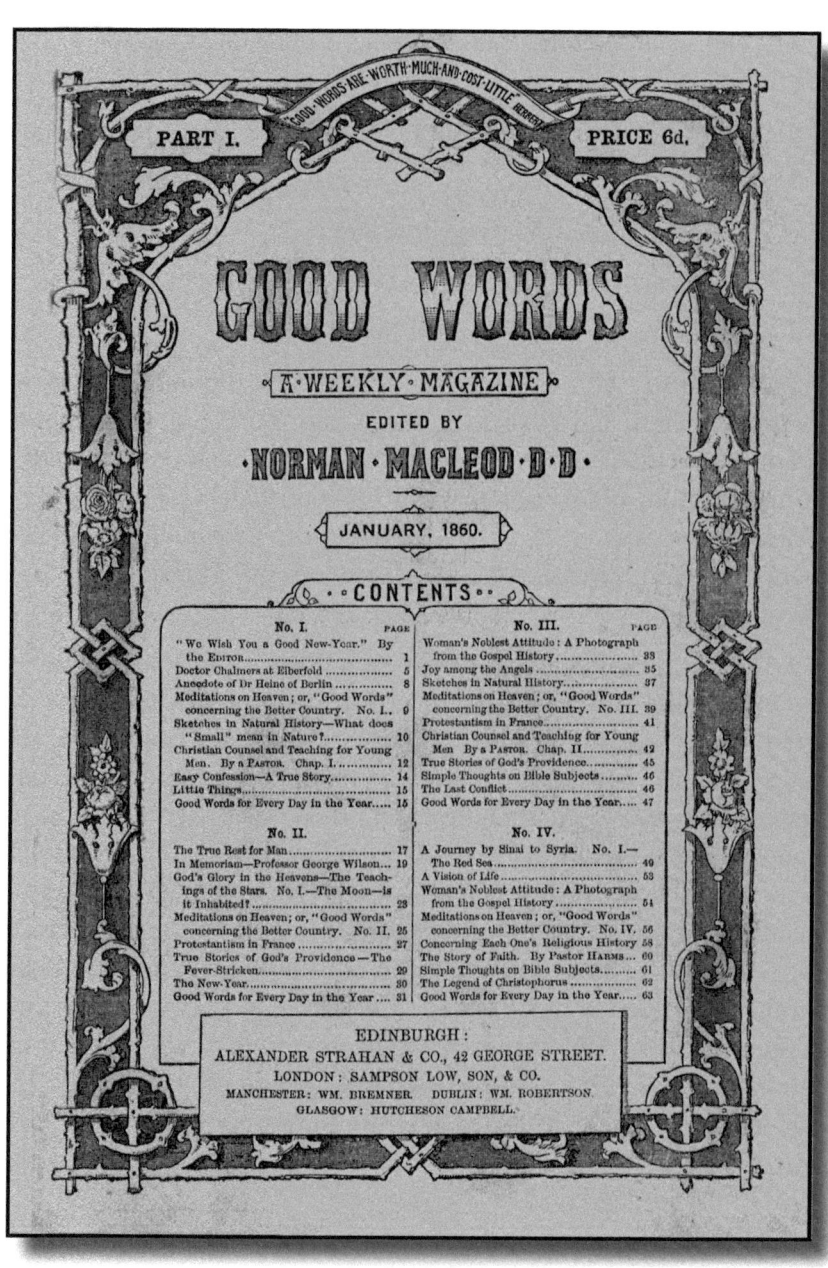

Cover of the 1st issue of Good Words (Jan. 1860)

There was no clear understanding of how electricity worked. There was no radio. Telegraphy had only just begun to flourish but the first trans-Atlantic cable had failed after the gutta-percha waterproofing had succumbed to the pressure of the ocean floor. Watt's steam engine had only just been successfully adapted for use at sea, the very first solely steam-powered ship to cross to America was a recent event. Brunel had only just built the massive steamer, SS Great Eastern, which ran on one steam driven propeller, two massive paddle wheels and six masts of sail. Oil had been discovered in America but the first shipment of black gold to Europe was only to occur that same year (1860). The atom was considered indivisible so the notion of atomic energy was decades in the future. Even the electron had yet to be discovered. Mendeleyev had not yet cracked the mysteries of the elements and drawn up his periodic table. Galaxies were all still considered to be nebulae inside our own Milky Way. Jules Verne hadn't even taken a fictional voyage in a balloon yet and was a year away from starting his first conversations with his mathematician cousin about outer space.

But that isn't to suggest that Leitch was living in the dark ages: far from it. The scientific community was in a state of excitement and agitation that would bring the world untold new discoveries over the next few decades.

Essay #1 - The Moon – Is it Inhabited?

William Leitch was far from the first to speculate on the nature of the moon. We have other written accounts which predate him by centuries, (such as the aforementioned Lucian) but his has to be one of the earliest accurate descriptions of what it would be like to stand on the lunar surface. In the space of a few short paragraphs Leitch revealed that he understood the impact of a vacuum on everything from light and sound, to ballistics and winged flight. This first essay ran for only two and a half pages but it set the stage for everything that was to come.

"In the survey we mean to take of the heavens as illustrative of God's glory, we shall first direct our attention to the moon, our nearest neighbour. The moon will form the first step in the ladder by which we shall attempt to scale those heights from which we may command the widest range of the marvelous works of the Almighty.

Although we cannot by searching find out God, although we are baffled in our attempts to comprehend the Absolute, still there are stepping stones across the abyss of space, which enable us to enlarge our view, and to form a juster conception of the Infinite and the Eternal."

In simple English what Leitch was saying was that he intended to take his readers on an imaginary journey to an outpost where the Earth's place in the Universe was plainly revealed. The metaphor of a ladder was one he would come back to in later essays. The choice of language was carefully refined, probably by Leitch, but perhaps by Macleod or even Strahan; to make it clear that this was a man of religion writing a secular description. Although we know that Macleod's name was prominently featured as the editor, Strahan was known to carry out the day-to-day editorial duties. However, in this instance it may have been Macleod monitoring his friend's first essay since it includes a quote from Wordsworth, who Macleod had befriended previously.

Leitch went on to discuss the possible arrangement of life on the moon, and explained how *"we can conceive of intellect united to a very different corporeal organisation."* This was a straight-forward acknowledgement that nature was not bound to position intelligence only in the human frame. The idea of alien life had been under discussion for decades, as previously noted, Christian Huygens' book *The Celestial Worlds Discovered* published in 1695 had discussed aliens to an almost heretical degree, which probably explained why Huygens didn't allow it to be published until after his death. We might assume that following Huygens with similar suggestions in 1860 would have been easier, but far from it. In fact the subject of the nature of life, intelligence and the eternal soul had been turned on its ear at almost the exact moment that Leitch was writing his essay. Just six weeks earlier a man named Charles Robert Darwin had just published a book entitled *"On the Origin of Species by Means of Natural Selection, or the Preservation of Favoured Races in the Struggle for Life"*. On June 3rd 1836 the HMS Beagle with its studious passenger had stopped at Cape Town. Darwin paid a visit to John Herschel where it is believed that they discussed biology, astronomy and cosmology. More than two decades later Herschel became good friends with Macleod and Strahan. But in the summer of

1860 Darwin's book would almost completely capsize the ship of literal scripture.

Leitch could scarcely have chosen a more contentious moment to enter the debate about man's place in the universe. One of his oldest friends, the Duke of Argyle who had been a member of his writing club at Glasgow, also joined the debate with his own highly public interpretation of Darwin's theory. The first adverts for Darwin's book had appeared just two weeks before Leitch met with Morris in Edinburgh. A week before Leitch's approval as Principal of Queens, Thomas Huxley had already taken sides and written his first approbation of Darwin for MacMillans. As we well know this fight was far from over and it would not be until April of 1860 before Darwin's book would be reviewed in places like the *Edinburgh Review*, a prestigious magazine which had previously included contributions by Leitch.

Despite all of this, in his very first essay Leitch tackled the big question of whether there might be life on the moon. In accordance with that analysis he had to explain to his readers exactly what the lunar environment was like.[4]

"Every possible test has been applied, but no trace whatever of air has been found in the moon. Eclipses and occultations have been watched with the utmost care, but all in vain; some of the tests are so delicate, that if there was an atmosphere capable of raising the mercury one-sixtieth of an inch in the barometer, it would have been detected. If there is an atmosphere after all, how evanescent it must be compared with ours, which raises the mercury to about thirty inches. Could we conceive living creatures to exist in the moon without air, how strange must be the conditions of life!

"Think how strange life must be in the moon without an atmospheric medium. Eternal silence must reign there. A huge rock may be precipitated from the lofty lunar cliffs, but no sound is heard—it falls noiselessly as a flock of wool. The inhabitants can converse only by signs. The musician in vain attempts to elicit sweet music from his stringed instrument; no note ever reaches the ear. Armies in battle array do not hear the boom of the cannon, though rifled arms, from the low trajectory of the ball, must acquire a fatal precision and range. No moving thing can live aloft; the eagle flaps its wings

against the rocks, and in vain attempts to rise. The balloon, instead of raising the car, crushes it with the weight of its imprisoned gas.

"Again, the inhabitants, having no atmosphere to shelter them from the sun and store up its heat, must recoil with terror from its fierce rays. During the long lunar day, the ground must become as burning marl, from which the scorched feet shrink with pain; during the equally long night, it must be colder than frozen mercury. No fuel will burn to mitigate the rigour of the cold, and none but the electric light can avail to dispel the darkness.

"Then as to light, how strange are the conditions! At noon-day the sky is as black as pitch, except in the region of the sun; and the stars shine out as at midnight. When the sun disappears in the horizon, darkness is as sudden as the darkness of an eclipse, or the extinguishing of a candle in a room. The inhabitants, on the shady side of a range of mountains, must be in almost total darkness, though the sun is above the horizon; and a room, lighted by windows in the roof, must be in the same predicament, except when the sun shines directly down."

This long extract presents many solid insights into Leitch's range of knowledge. His description of the lunar environment was uncannily accurate. He had evidently monitored everything from eclipses of distant planets by the moon, to the Newtonian possibilities of accurate flight in a vacuum, and the impact of a vacuum on both sound and light. [5]

He also made a throwaway comment about "electric light," an invention first demonstrated by Humphry Davy in 1809 but still relatively rare in the world of 1860; the Edison light bulb still being 20 years in the future. He already knew this topic well, having lectured on the subject of electricity while in Glasgow and Monimail.[6]

Evidently Leitch and his old schoolmate William Thomson, the soon-to-be Lord Kelvin, had also corresponded on the subject. In August 1849 a bolt of lightning had struck a farmhouse in Monimail with such peculiar results that Leitch would send samples of melted metal to Kelvin. His letter would trigger an investigation by Kelvin in the laboratories at Glasgow and further investigation by Leitch in Monimail. The letter and results appear in Kelvin's papers on electrostatics and magnetism.[7] (By a strange coincidence, decades later,

Lord Kelvin had a highly reliable assistant working for him; another graduate of Glasgow University, who would travel around the world assisting in making improvements to Kelvin's undersea cables. His name was also William Leitch, but he was evidently no relation.)

Essay #2 - The Moon's Invisible Side

In his second essay (*The Moon's Invisible Side*) Leitch tackled the issue of whether the far side of the moon could be reasonably expected to be exactly the same as the visible side. He mentions the libration of the moon which allows earthbound observers to see slightly more than half of the lunar surface, as the moon wobbles back and forth. Leitch explained how the libration had been used to create stereo images of the moon's surface. His knowledge of this almost certainly came from Brewster, who had invented the first portable stereoscope.

Leitch then discussed the new theory put forth by the respected Danish astronomer Peter Andreas Hansen which had proposed that the moon's near face was elevated towards the Earth to such a degree from the lunar centre of gravity that it would be like scaling a very tall mountain. Hansen's supposition briefly led to the thought that a lunar atmosphere might have gathered in the "low" ground on the far side and therefore life might be possible there. A reviewer in 1863 chastised Leitch for *"unreservedly adopting this (Hansen's) theory"* but a careful reading of Leitch's comments show no such endorsement. Quite the contrary, Leitch carefully examined the new information and implied that it had only provided sustenance to the advocates for lunar life. The rest of his comments were always phrased in such a way to ignite the reader's imagination.[8] Leitch then explained how magnificent it would be if only we could conduct a discourse with life on the moon. Remarkably he even postulated the notion that *"We could readily conceive a code of signals by which telegraphic communication could be carried on."* Today the word *telegraphic* implies radio, or possibly wired communication, but despite the rise of the commercial telegraph over the previous 30 years he must certainly have been thinking in terms of flashes of light or some other form of visual communication.[9]

Perhaps the most poetic part of this essay is when he fantasized about how an inhabitant of the lunar far side would know nothing of the Earth unless they were to venture into the "highlands" and cross the rim to the near side. *"What an astonishing spectacle must burst upon the view of the lunar tourist as soon as he fairly gets within the new hemisphere! He will see an immense blue orb hung up, immovably fixed in the heavens. It will appear fourteen times larger than the moon appears to us. But though immovably fixed in the heavens, wondrous activities will be discovered. It will exhibit in 28 days all the phases of the moon…its rapid rotation will also be a most notable object…and then the blue atmosphere will be going through incessant changes."*

Essay #3 - Lunar Landscape

The third essay appeared in the third week of March 1860 and covered the subject of the lunar landscape. The most obvious features to which he drew attention were the many craters visible from the Earth. He also talked about how finely tuned the climate is on the Earth and how with the slightest change, things would be very different. In keeping with that thought he said, *"A slight change in the constitution of the atmosphere, or in the alternation of night and day, would be fatal to many forms of life. Did a comet come into collision with our earth, so as to change its axis, new conditions, wholly destructive to a wide range of animal and vegetable life, would be introduced?"* [10]

The notion that a cometary impact would be catastrophic for life goes back at least as far as Edmund Halley, but Leitch specifically postulated in the context of how it might completely upend the fine balance of the ecosystem. This notion was still aggressively challenged over 100 years later when Luis Alvarez made his famous explanation for the demise of the dinosaurs.

Essay #4 - Discovery of the New Planet Vulcan

About a month later Leitch turned his attention to the announcement that a new planet had been discovered inside the orbit of Mercury. It was dubbed Vulcan and astronomers had spent many fruitless hours seeking its shadow passing in front of the sun. Always writing from the perspective of piety Leitch would deftly stitch together a logical platform to explain each new discovery. In this case it was a few months earlier that the great astronomer Le Verrier had calculated that the only explanation for the perturbations in Mercury's orbit, was the existence of an as-yet undiscovered planet. In December 1859 a French village doctor named Lescarbault sent Le Verrier notification that he had observed the missing planet several months earlier. Leitch reveled in the fact that this discovery, like the discovery of Neptune, was no accident but had been predicted by careful calculation. The whole notion of *a priori* discovery appealed to Leitch as an elegant proof of man's God-given intellect.[11]

Of course Lescarbault was subsequently proven to be mistaken and the mysterious planet *Vulcan* was consigned to the history of oddities, but even today such famous names as Alan Stern, the chief scientist on the *New Horizons* mission to Pluto, is known to have searched for small asteroids inside the orbit of Mercury. It would be Einstein who would ultimately explain the peculiar behavior of Mercury.

Around this time, in keeping with the esteem and high regard of his colleagues Leitch was given the degree of Doctor of Divinity by the University of Glasgow on May 7th 1860.[12] But it would be more than a year later that his nomination for the post of Principal was approved by the Church in Scotland setting the wheels in motion for his final emigration to Canada.

Essay #5 - The Approaching Total Eclipse of the Sun

In September of 1860 Leitch opened an essay in Good Words by discussing how the sun's power was bottled up in coal, falling water and wind. This explanation of our sources of energy had also been articulated by the anonymous author of the "Sketches in Natural History" column the previous February. That author also cited as the source of the idea *"...a philosopher of our day, young in years but mature in thought and observation (Professor William Thomson)"*.

It wouldn't be until 1862 that William Thomson and his colleague Peter Guthrie Tait would begin work on their seminal book *Treatise on Natural Philosophy*. They would first present a sample of their thoughts on the subject in Good Words, under the simple title "Energy". Many of the things stated by Thomson and Tait about the natural storehouse of solar energy had been articulated by Leitch in Good Words in 1860. Again this seems to suggest Leitch as the author of the short-lived "Sketches in Natural History" column.

Leitch's essay mainly concerned itself with eclipses and in this he compared the magnificence of a total solar eclipse with the spectacle of Niagara Falls. His description of Niagara is brief but detailed enough to suppose that he may have already visited Canada before this was written. His account of an eclipse is much more exhaustive and it places the reader firmly in that year of 1860 when astronomers like Leitch, Le Verrier and Hansen were still attempting to explain the coronasphere and solar prominences. Leitch demonstrated that he was willing to admit what he didn't understand, rather than simply making things up, and he often challenged those around him who were less fastidious.[13]

Essays #6 & #7 Comets – Their Nature and Design & Comets – Their History

Through the summer months of July and August Leitch turned his attention to the mysteries of comets. He would spread his comments over two months and two issues and skipped the June issue because on June 19th 1860 a new comet had been spotted by a French army officer, spurring this protracted discussion by Leitch.

This event firmly places the composition of Leitch's essays in the moment when they were first published. i.e. at the very least, his August essay was definitely composed in late June or early July 1860. In the July essay he discussed the subject of how comets' tails seemed to be pushed away from the sun, a process which was the source of some consternation amongst his peers.

This apparent resistive force, visible to anyone with a telescope, seemed to defy Newton's laws of gravitation; a notion which Leitch was not eager to accept. Papers were published about how the sun and planets might have magnetic or electric forces at work to push the tail away from the sun. Leitch made reference to how in Newton's day the theologians were distraught that the celestial machine might somehow be imperfect, but with a typically enlightened attitude he concluded, *"Suppose there is a resistive medium accounting for the destructive course of (comets) towards the sun, or that the law of gravitation requires to be modified by some new law, which may compromise the present order of things by introducing an element of decay are we to conclude that there is a defect of wisdom in the constitution of the celestial machine? Theologians have been too much led away from the idea of the solar system being a rigid machine, with unvarying adjustments, instead of a single phase of the mighty evolutions of the material universe. There is nothing fixed and rigid in the material world except the laws by which the all-wise Ruler governs it."*

To conclude his essay he turned his eye briefly to the subject of life on comets. He dismissed the ruminations of some of his contemporaries who insisted that life could exist anywhere that God ordained it. *"But is it right to make our notions of possibility the basis of a theory of God's providence? The question is one of probability, not of possibility. And if we are to proceed reverentially, we must argue from the known to the unknown, from our experience on this globe to what is probably the Divine procedure in worlds the physical condition of which are only partially revealed to us."*[14]

After discussing the good fortune of the Baron de Marguerit, who had discovered *Comet III 1860* by sheer luck while out smoking a cigar, Leitch turned his attention to the consequences of cometary impacts. He didn't specifically take sides with the few people like Halley who had cautioned of the dangers of comets, but he also didn't dismiss the danger out of hand. In all of these essays Leitch revealed that he was a free thinker.

"The comet played a most important part in the rude geological theories of those days (Halley's). Comets are in much disrepute among geologists at the present day, as accounting for the various convulsions and cataclysms indicated by the earth's crust. In the days of (William) Whiston however, the tails of comets were all important, as they could tilt up the axis of the earth, and produce deluges at will, to account for the various geological phenomena."

He continued, *"To soothe the alarm, mathematicians have calculated the chances of a collision, and have shown the probability of such a catastrophe. Still they have not been able to show the impossibility; and the popular mind is sometimes as much alarmed at the possible as at the probable.*

"Astronomers may give an assurance that a comet will come in collision with the earth only once in 280 million years, but then they can give no positive assurance that the one time may not be in our day."

Leitch concluded this particularly insightful essay by indirectly providing the entire premise for H.G. Wells science fiction story *"The Comet"* (written in 1897) when he discussed the prevailing notion that the comet's tail might contain harmful gasses, which if introduced to the Earth's atmosphere could produce unfortunate effects. He then went on to explain how the orbits of comets were generally out of the plane of the ecliptic and therefore were less likely to be the cause of devastation before summing up with this:

"We may regard a comet as a plumb line let down into the depths of space, to explore the nature of the currents, and the objects that may exist far beyond the reach of vision. The change in their constitution, and the disturbance of their orbits, may well tell us of the existence of worlds which the telescope may never reach. As the Lead of the mariner, with its adhesive surface, brings up unmistakable evidence of the nature of the bottom, so comets, when we understand them better, may bring interesting news of regions hitherto unexplored." [15]

Notes:

1. Day Of Rest, Strahan and Co, 1881

2. The Spectator, Oct. 27 1860

3. At the end of Leitch's first edition he quotes the following laws: Laws of Motion, Kepler's Laws, Law of Gravitation, Elements of Elliptical Motion, Secular Variations in Orbit, Laws of Light, and the erroneous Bode's Law and Kirkwood's Law.

4. Good Words, Strahan, Edinburgh, #2 January 1860

5. Leitch's brother John travelled extensively. He lived in London and attended King's College before retiring to an estate near Rothesay. He died in 1880 having never married. William Leitch's daughter Moncrieffe married Joseph Gordon Stuart of Fife, she died in 1913. Leitch's son John would earn degrees at Edinburgh and St. Andrews before settling into private practice as a medical doctor in Silloth Cumberland where he raised his five daughters and two sons, all who would become accomplished golfers. Like his father John would also die of heart failure while in his 40s. Grandson William Gladstone Leitch became a flour miller and would serve in World War I. He survived the conflict and lived until at least 1937 in London, but it is not known if he has any descendants. Principal William Leitch's parish records are at St. Andrews University while his granddaughter's golfing memorabilia is at St. Andrews Golfing Museum.

6. Men of Fife by M.F. Conolly, Inglis & Jack, Edinburgh, 1866

7. Reprints of Papers on Electrostatics and Magnetism by Sir William Thomson, Macmillan, London, 1872

8. Littel's The Living Age, Boston, June 1863

9. Good Words, Strahan, Edinburgh, #6 February 1860. The notion of visual communication with the moon or planets continued to be discussed well into the 20th century and it was only with the suppositions of Tesla and Marconi that different frequencies of the electromagnetic spectrum entered the conversation.

10. Good Words, Strahan, Edinburgh, #11 March 1860

11. Ibid. #15 April 1860

12. Globe, Toronto, May 7, 1860

13. Good Words. #19 - May 1860

14. Ibid. #30 - July 1860

15. Ibid. #33 - August 1860

Chapter 5

Essay #8 - The Sun – Its Work and Structure

In the autumn of 1860 Leitch tried to explain the workings of the sun to his readers. The theories of his time didn't include the notion of nuclear reactions and so he can be forgiven for not producing a modern analysis. As we have seen Mendeleyev's periodic table was still several years in the future and the atom was considered irreducible until the discovery of the electron almost forty years after that. However, even allowing for this, Leitch demonstrated clear understanding that almost all of the options for energy on planet Earth are derived from the sun's power. He discussed, wind, chemical and solar energy and even got into the subject of how there is an immutable amount of energy and matter in the universe. His peers had pondered how the sun seemed to be able to burn constantly without being replenished, and believing that this was in defiance of all known natural laws, they postulated that it was fuelled by the zodiacal light or a steady influx of comets. Variations of this theory survived for at least another 70 years and indeed were subscribed to by the Austrian Max Valier, who in 1930 became the first person to propel a manned vehicle with a liquid fuelled rocket. Leitch however wisely chose to reserve his own judgment on this hypothesis by stating *"we have reached the boundaries of knowledge and even speculation is silent."* [1]

Essays #9 & #10 The Structure of the Planets & The Nebulae

Just prior to his first departure for Canada Leitch took on the task of explaining how the solar system might have formed. The prevailing theory was attributed to Simon Pierre de Laplace who had suggested the nebular hypothesis. By proposing that the solar system might have formed from a gradually cooling whirlpool of dust and gas Laplace had thrown one more wrench into the preconceptions of Christian orthodoxy. Inevitably this had upset many of Leitch's learned colleagues in the Church but once again Leitch took a level-headed approach. In his mind the Laplace theory was elegant, but it was still unproven; and even if it proved to be true he saw nothing within the theory that diminished his belief in an intelligent hand at work. Indeed, Leitch chastised his fellow theologians for staking the existence of God on the refutation of the Laplace theory. *"Instead of attacking the scientific theory, the proper attitude is to deny the religious inference. Instead of denouncing the theory as atheistical, the only tenable position is to show that, though granted, it would not warrant the atheistical deduction.*

"Granting that the solar system was developed from a nebulous mist according to the rigid laws of mechanism, the question at once arises; who endowed the atoms of this mist with such properties and susceptibilities, as to form worlds, and plants and animals?" [2]

It is a question that still perplexes the brightest minds alive today. Indeed, many physicists have asserted that even a minor change to any of the basic forces in our universe would preclude our very existence. Leitch could not have known this but nevertheless he was a man guided by acceptance of the facts as they emerged. He was willing to accept that there were correlations between the design of the planets and the stars that showed obedience to some set of universal rules. The fact that he ascribed the existence of those rules to intelligent purpose was simply a mark of his circumstances. He even seems to have had some inkling of these natural tendencies from both the macroscopic and microscopic worlds, comparing tree rings to spiral galaxies, and in some small way presaging the findings of Mandelbrot and others in the years ahead.

Leitch would leave for Canada in October of 1860. He brought with him a library of more than 600 books. This collection is now dispersed into the collection of Queen's University library but it holds many insights into Leitch and his numerous interests. Of course there are works by Newton and Laplace, but there are also books on botany, physiology, geology, mathematics and many books of religion.

Shortly after Leitch left Scotland an essay appeared in Good Words entitled "Auroras". It is another article which is unattributed, but again it reads a lot like Leitch's work. It is crystal clear that it was written by a scientist who studied the sky but it doesn't quite fit neatly into a work of astronomy. It is of particular interest to the story presented here because the comprehension of auroras and the Earth's magnetic field would ultimately be the main stimulus for Canada's space program and its position as a world leader in communications in the next century. Without any real understanding of how the sun worked, the author of this article was left to speculate on the cause of auroras and the interference they caused in the telegraphic communications cables of the time.

Arrival in Canada - 1860

Leitch arrived in Kingston Ontario on 19th October 1860. His steamer, the *Nova Scotian*, had made the journey from Liverpool to Montreal. Just as he was leaving the ship he was introduced to Fennings Hope, the Clerk Assistant of the Legislative Council of Canada. Hope immediately took a liking to Leitch and would later write a biography of him for William Notman's book of photographs entitled *"Portraits of British North Americans"*.[3]

Just before Leitch had arrived by Canadian steamer into Halifax Nova Scotia the giant Brunel steamer *SS Great Eastern* had passed through with great fanfare on its first voyage to the Americas. The ill-fated super-ship would later become instrumental in laying Lord Kelvin's undersea cables, bringing communications to the world.

After settling into his new role at Queens College in Kingston Leitch took on a multitude of tasks. He created a new faculty of law, he worked to get the observatory and its telescope into more permanent surroundings. He founded the Botanical Society of Canada. And perhaps most importantly he took on the voice of advocate for a free and expansive system of further education in Canada.

Leitch's first year in Canada was to be probational before he would agree to accept the post full-time. As has been previously noted his writings seem to imply that he might have been to Canada previously, but so far no solid evidence has come to light to support this. His earlier descriptions of Canadian forest fires, or of the tourists at Niagara Falls, may have come from hearsay. We also know that he owned a copy of William Hunter's *Guide to Niagara Falls* in his library. Regardless, we know with certainty that he arrived at Queen's College and was inaugurated with some fanfare into the role of Principal of that body on November 8th 1860.[4]

Canada in 1860 was still very firmly a part of Victoria's empire. In 1798 King George III of England had made a generous endowment of hundreds of thousands of acres of royal land to the people of Canada with the express purpose that the revenue generated from that land be used to create a robust system of education for all Canadians. By the 1830s most of the land had been sold and a vast sum of money was being held for the purpose of setting up seats of higher learning. When Leitch had left for Canada in the autumn of 1860 he understood that he would be taking on the leading role at the Queen's College, which had been founded in the 1840s by the Presbyterians of the Church of Scotland. It was one of many such colleges, each established and run by different Christian denominations. The role of Principal had remained vacant at Queens for fourteen years and so Leitch's arrival was anticipated with some enthusiasm.

It seems extremely unlikely that Leitch would have been dispatched to Kingston from Glasgow without having at least some insight into the political situation at Queens. However, one could be forgiven for assuming otherwise. As it happens, the challenge was one of some considerable proportion. The aforementioned endowment handed over by the monarch in 1798 had been the cause of significant agitation for the better part of three decades. When Leitch arrived he walked into a tangled web of political intrigue involving money, politics and power. Leitch was being paid $200 a month in his role as Principal and yet the purse strings were being held so tightly that he couldn't even get travel expenses. [5]

The University Question

When George III had gifted the land to the people of Canada it was specifically to raise money to establish a system of schools and seats of higher learning. In an echo of what had been happening in Scotland and England it was understandable that the various denominational schools, like Queens, assumed that they would be given a piece of that endowment's earnings to operate their institutions. This same assumption was duplicated by the authorities at schools run by the Methodists, Catholics, Baptists, Anglicans and the Presbyterians (represented primarily by Queens in Kingston.) However, that was not the case.

Several acts of the government had been passed during the intervening years to resolve this dilemma, and when Leitch arrived he tackled the problem with his usual articulate intelligence. In his mind he saw a school system akin to that which he had seen in

Glasgow and Oxford. There should be one university for examining and bestowing diplomas and degrees; and a series of satellite colleges to teach the curriculum to the students. In his mind all of the key subjects could be standardized with the exception of Divinity which he freely acknowledged might take on a different shape in each school. This was his solution to what must have seemed a fairly straight-forward problem. [6]

Before he took on the fight for the future of Canada's educational system he first set about establishing the Botanical Society of Canada. The newspapers back in Scotland reported his progress.

"A Botanical society has been established in Kingston Canada West, having for its object the thorough investigation of the Canadian Flora, as well as the prosecution of the science of botany in general. The Society seems to have a similar constitution to that of the Edinburgh Botanical Society. At the first meeting held at Kingston on Friday 7th December last there was a large attendance of gentlemen from all parts of the country; and after addresses had been delivered by the Rev. Principal Leitch, Professor George Lawson and Professor Litchfield, there were upwards of 100 names enrolled as members of the Society."

Leitch became the society's first president. His interest in plants undoubtedly came from his work with Thomas Brown in Scotland. He wrote at least one paper on the impact of insects on Pear Trees. Combined with his solid understanding of bees he must have made a persuasive teacher on this subject. [7] [8] He also became President of the Medical Faculty and appointed a fellow Scot named Alexander Campbell to run the new Faculty of Law. Campbell was partner in a Kingston law firm with the town's most distinguished resident, John A. Macdonald, and would later become Lt. Gov. of Ontario and a Canadian Senator.

Having barely been on the job for four months the observer is left to assume that Leitch must have looked at the College's finances and determined that he needed to delve into the endowment issue with all of his talents, much as he had done back in Fife a decade earlier. On Wednesday March 6th 1861 the largest gathering ever assembled of the residents of Kingston took place in the City Hall to attend to what seems to have been the general opinion that Queen's College was being swindled of its share of the King's endowment. The main speaker that night was Leitch. His speech would later be printed and disseminated and was spread over 15 pages. Having barely arrived in his new country he had staked his fortunes on one side of a fight which would not be resolved in his lifetime and, some people at the time believed, contributed to his early death. [9]

The Observatory in Kingston (ca. 1900)

Essay# 11 - The Kingston Observatory and Alvan Clark - 1861

But before the fight for educational rights consumed him, Leitch quickly returned to his great love, astronomy. One of the first things he resolved was the transfer of control of the Kingston Observatory to the College. The observatory had been announced in 1855 with the express purpose of providing a world-class facility for the students. It had been built with subscriptions and was to be furnished with optics from one of the best manufacturers in the United States. In 1861 Leitch arranged for the building of a more permanent brick structure to house these instruments.

In the summer of 1861 Leitch returned to Scotland. His children remained there and demanded his attention. The journey took him via Albany New York and Cambridge Massachusetts. In Albany he visited astronomer Ormsby M. Mitchel at the Dudley Observatory who was known for his many magazines and books on astronomy. Mitchel had just become a Brig. General in the Union Army. The Dudley facility would later move locations, but it still claims today that it is the oldest continuously operating observatory in the United States. In Cambridge Leitch visited Alvan Clark, one of the most accomplished makers of optical instruments in the world. Leitch was presumably there to discuss the capabilities of the Kingston equipment which Clark had provided. He found Clark putting the finishing touches to what would be the famed 18.5" Dearborn telescope. Leitch was so impressed by Clark's work that he wrote an essay about him for publication in *Good Words*. Of incidental interest was the fact that the American Civil War was in its first few months when Leitch arrived in Boston. The enthusiasm of the rallying troops caught his attention. He even cited an extract from a song on everyone's lips; the lyrics were those penned by Francis Scott Key and ultimately destined to become the American national anthem. However, Leitch was evidently quietly saddened by the spectacle and stated, *"It was gratifying to find one man engaged in the peaceful pursuits of science."* [10] Unlike Clark, Ormsby Mitchel had immediately abandoned astronomy and joined the fray. In 1862 he led the force which took Huntsville Alabama; later the home of America's rocket program.

Leaving Clark to his lenses and polishing, and Mitchel to his battle plans, Leitch embarked on a steamer for Scotland. It is at this time that we are obliged to carefully consider the dates and times to allow us to make any assumptions about his next, and possibly most important, scholarly work.

Alvan Clark (centre) with his sons.

Notes:

1. Good Words, Strahan, Edinburgh, #37 September 1860

2. Ibid. #40 October 1860

3. Portraits of British North Americans by William Notman, Montreal, 1865

4. Globe, Toronto, Nov. 10, 1860

5. Documentary History of Education in Upper Canada by J. George Hodgins, Vol 18, Cameron, Toronto, 1907

6. Globe, Toronto, Apr. 14 1849

7. The Journal of the Boards and Arts and Manufactures for Upper Canada, Vol.1, 1861

8. On The Selandria Aethiops and its destructive effects on Pear Trees by William Leitch, The Canadian Naturalist and Geologist and Proceedings of the Natural History Society of Montreal Volume 8, Dawson Brothers, Montreal 1863

9. University Reform Report of the Resolutions Adopted at a Great Public Meeting of the Inhabitants of Kingston, 1861

10. American Telescopes and Astronomers, Good Words, Strahan, London, July, 1861

Chapter 6

Over 18 months after publishing his book Charles Darwin would finally be challenged in the pages of Good Words; but not by Leitch. The article was written by none other than David Brewster. Brewster was now one of the world's foremost authorities on optics, which was yet another interest of Leitch's.[1] Brewster, despite believing in alien life along with many other scientists, couldn't accept the notion that all life had come from a single primordial form. The main reason for including this point is to demonstrate that *Good Words* had become one of the most widely read and respected forums for scientific discourse of its time. Other articles in 1862 were written by Charles Piazzi Smith, the Astronomer Royal for Scotland who wrote "Above the Clouds" and "Time and its Measurement"; James David Forbes, FRS, discussed "Glaciers"; James Glaisher took the readers on a beautifully described scientific balloon voyage into the stratosphere, and by 1863 Sir John Herschel tackled the meteoric theory of solar power and the study of volcanoes. Leitch was in estimable company.

Essay #12 - A Journey Through Space - 1861

We know that Leitch arrived in Boston after the beginning of hostilities in early 1861. We also know he then subsequently embarked for Scotland. In his essay on Clark he talks about being on the steamer so we can perhaps assume that he wrote his treatise on Clark during the crossing. It appeared in *Good Words* in the July 1861 issue (Good Words was now a monthly) and it was followed in August by an explanation of his work with bees.[2] This would suggest that these works predated his next essay which would return him to the world of astronomy and appeared in *Good Words* in the September issue. Perhaps he also wrote it during the crossing, or perhaps it was composed before leaving Canada. All we can know for certain is that it was written after the summer comet of 1860 but wasn't ready for publication that year, but it was finished by the time he returned to Glasgow in the autumn of 1861. It seems improbable that he had time to write it during his first few months of tenure at Queens, but not impossible. The timing couldn't have been better. The demand for *Good Words* was now flying high. The noted review periodical *The Bookseller* writing in September 1861 on the state of the competition between the hoard of magazines being published stated, *"...the only one that is progressing is Good Words, which appears to devour all others."* [3]

Leitch's next essay in *Good Words* would be entitled *"A Journey Through Space"* and this time he was to outdo himself.[4] In this essay, published in the first week of September 1861, he pursued his earlier metaphor of scaling the rungs of an infinite ladder to get a better view of our solar system. It was his twelfth essay on astronomy but it was destined to be the opening chapter in his forthcoming book. In six thousand words he took his readers on an imaginary tour of the solar system. In just six pages he dazzled his readers with an astonishing array of controversial ideas.

Leitch began by making a quick summary of how important the spirit of man is, even when faced with the immensity of the cosmos. His argument, couched in wholly Christian terms, was that despite the work of astronomy constantly revealing humans to be smaller and smaller, the result in his mind at least, was how important this made us. This same notion was articulated more than a century later by noted science writer Arthur C. Clarke who submitted that if humans are alone in the universe, *or not*; the confirmation of either fact would be equally astounding.

Using his most eloquent prose Leitch then decided to elevate his reader to a place outside of the usual observation post of planet Earth and take them on a tour of the universe.

The Rocket for Space Flight - 1861

"Though the facts and deductions of astronomy sufficiently bring out the immensity of the universe, as contrasted with our world, still it is difficult to realise the truth; our thoughts will obstinately cling to our globe, and the images of grandeur will still be, our terrestrial seas and mountains. Let us, however, attempt to escape from the narrow confines of our globe, and see it, as others see it, from a different point of view. Let us take a nearer survey of other orbs and systems, and see what impressions they produce, as compared with that received from the platform of the earth.

"But what vehicle can we avail ourselves of for our excursion? Must we be altogether dependent on the fairy wings of imagina-

tion, or can we derive aid from some less ethereal agencies? It was long the fond wish of man to soar above this terrestrial scene, and visit other planets. In the infancy of physical science, it was hoped that some discovery might be made that would enable us to emancipate ourselves from the bondage of gravity, and, at least, pay a visit to our neighbour the moon. The poor attempts of the aeronaut have shown the hopelessness of the enterprise. The success of his achievement depends on the buoyant power of the atmosphere, but the atmosphere extends only a few miles above the earth, and its action cannot reach beyond its own limits.

"The only machine, independent of the atmosphere, we can conceive of, would be one on the principle of the rocket. The rocket rises in the air, not from the resistance offered by the atmosphere to its fiery stream, but from the internal reaction. The velocity would, indeed, be greater in a vacuum than in the atmosphere, and could we dispense with the comfort of breathing air, we might, with such a machine, transcend the boundaries of our globe, and visit other orbs.

"Instead, however, of torturing our imagination to conceive of a rocket device, which would eclipse the performances of all flying machines, let us take one of nature's rockets as the material aid to our imaginary flight. Let us follow the course of some comet in its wanderings across our system. A rocket, held fast, with its fiery stream directed against a strong wind, very well represents the telescopic appearance of a comet, when in the neighbourhood of the sun. The luminous particles shoot out from the nucleus of the comet, precisely as the sparks issue from the rocket-tube, and they are thrown back as a strong wind would throw back the fiery stream of the rocket."

This remarkable opening volley by Leitch seems at first blush to overthrow a century of space history. The paragraph highlighted above in bold type offers no opportunity for misinterpretation. It is succinct and self explanatory. Leitch knew that the reaction forces implicit in the rocket made it the only engine which would work in space, *and it would work better in space.*

These two facts are in direct contravention of the accepted history of this subject.

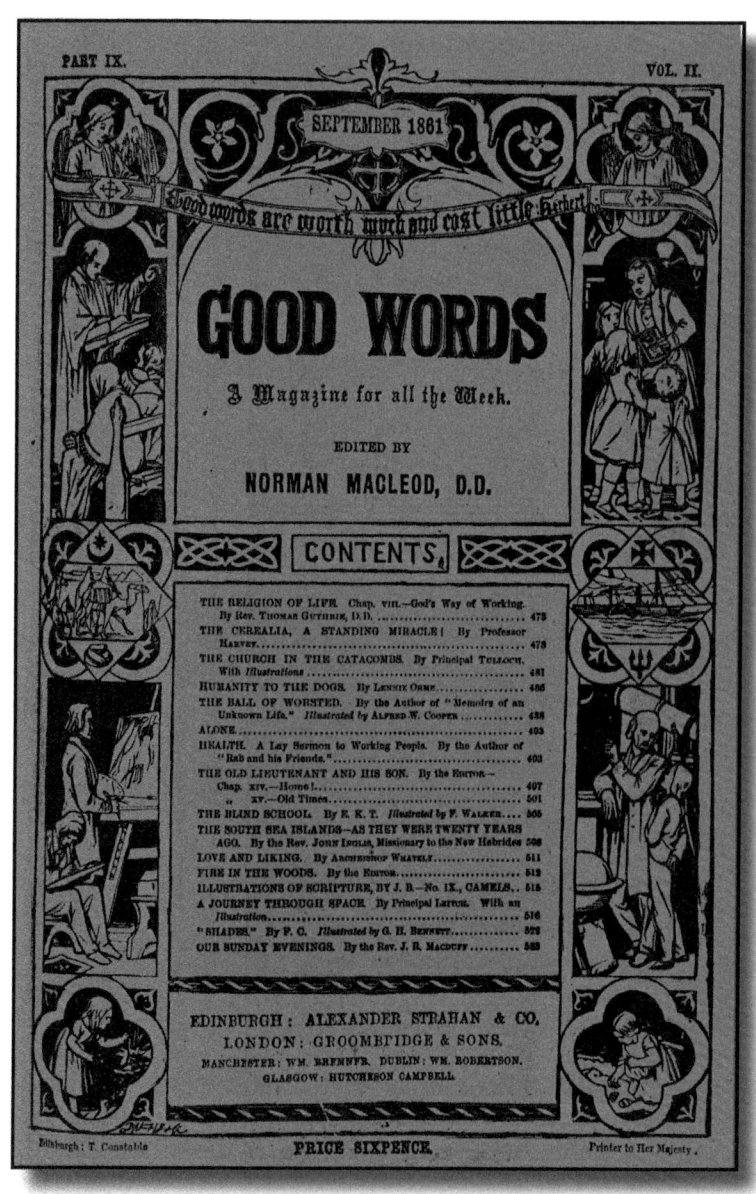

Cover of the "rocket" issue of Good Words
(Sept. 1861)

Newton's Laws

There are two slightly different but similar reasons why a rocket's velocity is "greater in a vacuum than in the atmosphere". They both involve the impeding nature of air molecules.

If a rocket accelerates away from the earth, the air around it gradually gets thinner and becomes less of an encumbrance to its flight, but if the rocket continues to get faster it has to push more air out of the way. Eventually a balance is achieved where the density of the air and the accelerating rocket hit a point where the two are at their most contentious. In the parlance of NASA this moment is called "Max Q" or maximum aerodynamic pressure. From that point onwards the air pressure declines and the rocket continues into the vacuum of space. All the time becoming more efficient because it is no longer being impeded by the air.

It seems that Leitch was clearly aware of the effect of a vacuum on things in flight. He states as much in his essay about the moon. *"Think how strange life must be in the moon without an atmospheric medium… Armies in battle array do not hear the boom of the cannon, though rifled arms, from the low trajectory of the ball, must acquire a fatal precision and range."* His last word is critical, "range". From this we can see that he understood that a vacuum removes this hindrance to forward flight. There is no "Max Q" in space.

The other reason that a vacuum improves the efficiency of the rocket is more subtle, but also important. The whole notion of a "reaction" engine is that if you throw mass in one direction it causes movement in the opposite direction. In the case of the rocket the "mass" is hot gas coming out of the rocket nozzle. The faster you can push the hot gas out of the nozzle the faster your rocket moves in the opposite direction. However, when you fire a rocket in the atmosphere at sea level the hot gas has to fight the air pressure around the nozzle. As the rocket ascends, this pressure declines steadily until the rocket reaches space. Evidence of this can be seen during any major rocket launch; as the rocket gets higher the plume of its exhaust widens as the air pressure declines.

This is where the argument began for those who did not understand Newton's third law. The rocket naysayers argued that without air pressure to *push against*, the rocket wouldn't move. Therefore it couldn't be used in space.

However, we can see from his own words that Leitch clearly understood that this was not the case. "The rocket rises in the air, **not from the resistance offered by the atmosphere to its fiery stream**, but from the internal reaction."

Quite clearly Leitch spells out the explanation for why the lack of air in space is a boon not an impediment. In space there is no air to inhibit the rapid flow of the exhaust. It would take another six decades before Robert Goddard would prove this through experimentation.

Leitch's description is worded in such a way to suggest that this was not even an important revelation to him. With very little effort the reader could be persuaded that this was something Leitch had discussed previously and had simply dismissed as so improbable that he gave it very little further consideration. However his comment about "torturing his imagination" suggests otherwise. His knowledge of projectiles is implicit.

Earlier Efforts

Certainly there were earlier references to rocket flight, but the few that exist were apocryphal or satirical, or simply written by people who had no understanding of why a rocket might work. The oft-cited Chinese adventurer Wan Hu, with his rocket propelled chair, is almost certainly an example of complete fiction; very likely created in the early 20th century. Then there is the French satirist Cyrano De Bergerac. Born in 1619, Cyrano was a soldier in the French army who was well known for his outrageous tales. It is of interest to note that Cyrano wrote two hilarious satires about space travel. *The Comical History of the States and Empires of the Moon* (1656) and *The Comical History of the States and Empires of the Sun* (1662). The single most significant contribution made by Cyrano would be his proposal to actually use rockets as a method of propulsion to send his protagonist (himself) to the moon. However, Cyrano did not hit upon this idea due to any exceptional scientific awareness or belief in its viability. An alternative method that he proposed was to fill bottles with dew, and then simply by strapping the bottles to his belt, he was miraculously lifted skyward by the inevitable action of the sunlight on the dew. Isaac Newton's revelations were still 25 years in the future.

There can be no such misinterpretation of Leitch's words. He not only very clearly understood how a vacuum was an impediment to any form of winged flight, but as an expert astronomer, he was also well-versed in Newtonian physics. Most remarkably he showed a very clear understanding of the dynamics of gasses in a vacuum and in an atmosphere. He dismissed the fumblings of the "aeronaut" (at that time a reference to balloonists, as this was 42 years before the Wright Brothers) and concluded that the rocket would far outstrip any other flying vehicle.

One of Cyrano's other methods for spaceflight was strapping bottles of dew to his belt.

Notes:

1. American Telescopes & Astronomers, Good Words, Strahan, London, 1861

2. Bees and the Art of Queen-making by William Leitch, Good Words, Strahan, Edinburgh, August, 1861

3. Alexander Strahan Victorian Publisher, by Patricia Thomas Srebrnik, Univ. Of Michigan Press, 1986

4. A Journey Through Space, Good Words, Strahan, Edinburgh, Sept. 1861

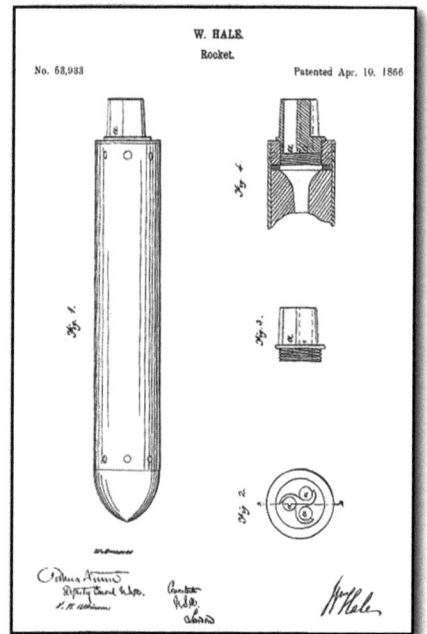 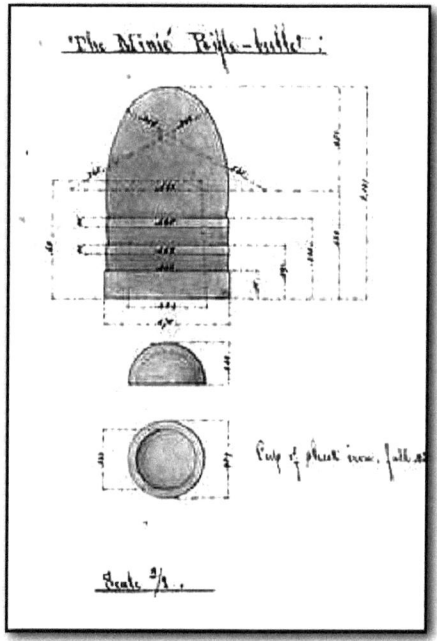

An 1866 patent for Hale's rocket showing the rotation vents and an 1850 sketch of a Minié bullet showing the soft skirts.

Chapter 7

Context

So what possible sequence of events could have led this gentle cleric from the Scottish countryside to postulate such a thing?

Perhaps we must look to his essay in August of 1860 for a clue. During that summer the subject of comets was on the top of his agenda. When Marguerit spotted a new comet intruding into the summer sky Leitch and just about every other astronomer studied it and filed reports. Leitch even mentioned that the stories of the previous bright comet in 1811 were fading into memory and the few remaining living observers were not proving to be very reliable witnesses. The most contentious debate raging around the observation of *Comet III 1860* was a reasonable explanation for the comet's tail. Leitch likens it to the sparks flying from a rocket. He clearly understood that it was not "flaming" and was comprised of extremely tenuous filaments of gas and dust, but the metaphor of the rocket may have sat in the back of his mind. He had also only just come from Boston where everyone was singing about the notorious bombardment of Baltimore by British Congreve rockets.

Certainly we know that his understanding of ballistics was enough for him to lecture on the subject. The Minié rifle had been introduced just a few years earlier when Leitch was still in Scotland. Interestingly this revolutionary weapon had also undergone recent improvements to allow breech loading, with the patent filed by someone named Leitche. Both Minié bullets and Hale rockets were manufactured at the same time at the Woolwich Arsenal.[1] [2]

This would all seem to suggest that Leitch may not have been thinking of firework rockets, but possibly the newly devised spin-stabilised Hale rocket.

We know that Leitch was not the first to understand the Newtonian principles at work in a rocket, but he seems to have been the first, that we know of, who had the background in mathematics, astronomy, ballistics, and natural philosophy to correctly articulate how a rocket would behave in space; and most importantly that it would work better in space. His grasp of the exact nature of the vacuum of space is as important as his knowledge of orbital mechanics and cosmology. His training in all of these disciplines are plainly evident. Undeniably the earliest reference for an informed scientific assessment of the rocket as the perfect engine for space travel currently resides with Leitch.

What are the implications of this? Is it possible that his writing landed on the doorstep of Tsiolkovsky or Goddard? We may never know, but we can certainly not preclude such a supposition. In fact we have solid evidence that Leitch's essays did indeed remain in print until at least 1910, long after Goddard had his "vision", without so much as a word altered. This also made them contemporaneous with Tsiolkovsky and Fyodorov (for more on this see page 182).

October 27, 1860.]

THE SPECTATOR. 1031

Messrs. A. Strahan and Co., Edinburgh, announce as forthcoming, "God's Glory in the Heavens; or, something of the Wonders of Astronomy," by Wm. Leitch, D.D., Principal of Queen's College, Canada; and "The Story of the Mission Fields; how they were planted, and how they have prospered," by the Reverend Thomas Smith, M.A., Edinburgh.

The first advert for Leitch's book.
Two years before publication.

Earlier References and Later Editions

The first modern reference to Leitch's insight to which we can certainly point is the book published in 1953 by the Chairman of the British Interplanetary Society, Kenneth W. Gatland entitled *"Space Travel"*. Written four years before Sputnik, Gatland and his co-author Anthony Kunesch actually cite the exact quote about the rocket, but were unaware of when or to whom it should be attributed. Gatland writes, *"It is indeed surprising to find in a book 'Half Hours in Air and Sky', published by James Nisbet and Co in 1899 an accurate definition of rocket theory...When it is considered that even today the principle of rocket motion is constantly being misinterpreted, the significance of this short paragraph written so long ago by an unknown author, is truly remarkable."* [3]

Gatland knew that what he was reading was extremely important but he did not have the advantage of being able to "Google" the text to investigate further. In fact the book he cited was a later text for juveniles that included Leitch's essay about a journey through space without any attribution at all.

Noted space historian and artist Ron Miller also seems to have come across Gatland's "find", which he mentions in his 1993 book *The Dream Machines*. Miller observed, *"The unknown author of these lines shows an appreciation and understanding of the principles of rocket flight unusual even among scientists of his day."* Miller had the knowledge to also appreciate that this may have been written much earlier than 1899, and stated as much in his book. [4]

This begs the question, how did Leitch's name become disengaged from his own writing? It seems that around 1859 his publisher Alexander Strahan was in partnership with a certain William Isbister. [5] Good Words had been their first major venture and Leitch's book was probably one of the first to appear on the Strahan imprint. After Leitch died, Strahan was called to task by his investors. It seems that Strahan had built a substantial following for not only Good Words but also several other magazines he had been publishing. To pursue such a large publishing concern he had been obliged to bring his printers into the business as silent partners, who were effectively operating as his bank.

By 1872 Strahan had become one of the largest publishers in the English speaking world and had secured publishing rights to such notables as Alfred Tennyson and William Gladstone, but despite massive circulation for Good Words the company was never really profitable, in part because Strahan was overly generous with his payments to authors. Tennyson had even talked Strahan *down* from his original offer.[6] That same year Good Words was subjected to a backlash amongst the evangelical sector of the church. On one occasion people had rounded up every copy of the magazine they could find and burned them; which might partly explain why it is so difficult to locate individual copies of the magazine today. In 1872 the cabal of printers and other publishers which had been floating

Strahan's efforts removed him and handed the company over briefly to William Isbister. The collateral which Strahan had used to secure the investors' support were the copyrights to Good Words. Isbister continued to publish Leitch's book until a complex series of share-swaps and takeovers brought an end to the original Strahan company by May of 1878.[7] However, just before the last of the shutdowns Isbister had begun the series of juvenile science books he called the *Half Hour Library of Travel, Nature and Science for Young Readers*. The volume on "Air and Sky" had first appeared in October 1877.[8]

Alexander Strahan (ca. 1860)

Because Isbister's new owners managed to retain the copyrights to some of the works Strahan had published, from 1878 onwards they stopped printing Leitch's works in full and simply appropriated several of the essays including "*A Journey Through Space*" and added them to their *Half Hour Library* without attribution. Perhaps most remarkable is the fact that it then seems to have never gone out of print, with editions known to have been published in 1878, 1880, 1882 and1886 before the rights were then turned over to the James Nisbet Company who printed it again in 1896, 1899 and 1903. It was exported to Australia and as late as 1907 was still being advertised. One such advert claimed that it had sold half a million copies. [9]

The fact that we can now confirm that the true date for this text places it before even Jules Verne's famous story puts its significance on an entirely different level.

Ironically it would be a similar circumstance which in the 1890s would lead to the works of George Griffith sliding into obscurity. His publisher also went into bankruptcy and took his copyrights with it. Griffith would invent the space launch countdown in his 1897 short story *The Great Crellin Comet* but no one would remember this until more than a century later, instead giving the credit to the 1929 Fritz Lang movie *Frau Im Mond*.[10]

HALF-HOUR LIBRARY OF TRAVEL, NATURE AND SCIENCE (The). In attractive binding, and containing nearly one hundred illustrations in each volume. Crown 8vo, 2s. 6d. each. Gilt edges, 3s.; postage 5d.—1. Half-Hours on the Quarter Deck: The Spanish Armada to Sir Cloudesley Shorel, 1670. 2. Half-Hours in Air and Sky: The Marvels of the Universe. 3. Half-Hours in Field and Forest, by Rev. J. G. Wood, M.A. 4. Half-Hours in the Deep: The Nature and Wealth of the Sea. 5. Half-Hours in the Tiny World: The Wonders of Insect Life. 6. Half-Hours in the Holy Land, by Norman Macleod, D.D. 7. Half-Hours in Many Lands: Arctic, Torrid and Temperate. 8. Half-Hours with a Naturalist: Rambles near the Seashore, by the Rev. J. G. Wood, M.A. 9. Half-Hours in the Far North: Life Amid Ice and Snow. 10. Half-Hours in the Far South: The People and Scenery of the Tropics. 11. Half-Hours in the Far East: Among the People and Wonders of India. 12. Half-Hours in Woods and Wilds: Adventures of Sport and Travel. 13. Half-Hours Underground: Volcanoes, Mines and Caves. 14. Half-Hours at Sea: Stories of Voyage, Adventure and Wreck. 15. Half-Hours in the Wide West: Over Mountains. 16. Half-Hours in Early Naval Adventure, wtih many illustrations.
Reduced to 2s. Each, Postage 4d.
Sales Half a Million.

Advert from 1902 showing how Macleod's name was still attached to his work but Leitch's is absent.

Notes:

1. The National Magazine, Vol 1, National Magazine Company, London, 1857

2. The International Exhibition of 1862, Vol 2, Clay, Son and Taylor, 1862

3. Space Travel by Gatland and Kunesch, Philosophical Library, New York, 1953

4. The Dream Machines by Ron Miller, Krieger, Malabar FL, 1993

5. Bookseller Gazette, London, Feb. 26, 1861

6. Alexander Strahan Victorian Publisher, by Patricia Thomas Srebrnik, Univ. of Mich Press, 1986

7. The Story of my Life by Augustus Hare, Dodd Mead, 1901

8. The Spectator, London, Oct. 6, 1877

9. The Age, Melbourne, Victoria, Dec. 20, 1902; The Argus Melbourne Aug. 3, 1907

10. The World Peril of 1910, George Griffith, Apogee Books, 2006

Chapter 8

Touring the Solar System on a Comet

The revelation of the rocket and its subsequent substitution by a comet all appear in the first two pages of Leitch's essay. But his insights into astronomy were barely started. In this extract we are shown exactly why Leitch thought the comet made a suitable imaginary space vehicle.

"The densest comet would afford but insecure footing to beings of almost spiritual essence, as the matter of which it is composed must be so light that the atmosphere of our earth is as lead compared to it. But we shall overlook this difficulty, and venture, in thought, to follow the fortunes of some cometic wanderer. The difficulty of reaching some suitable comet is lessened by the consideration that the comet may come to us.

"The great advantage of the comet, as a convenient vehicle for an excursion, is that it gives near, as well as extensive views of the system. Most comets rise above the plane of the solar system, so that we may have a clear view of the relation of one planet to the other. Then, again, let us consider the rate at which the comet travels. This is by no means an equable one. Sometimes it moves so slowly, that a child might keep up with it; at another, it speeds round with lightning velocity. It is like a coach going down a declivity without a drag. It increases its velocity till it comes to the bottom of the hill, and the momentum acquired carries it up the opposite side, till it gradually slackens and assumes a snail's pace. The comet approaching the sun is going downhill, and when it reaches the nearest point it wheels round, and then ascends till its speed is gradually arrested. It is reined in by the sun, from which there are invisible lines of force dragging it back; and, if its momentum be not too great, it is effectually checked and brought back to pursue its former course. Most frequently, however, its course is so impetuous that all the strength of the sun, in reining back, avails nothing. It breaks loose, like a fiery steed from its master; speeds off into space, and is heard of no more."

Having boarded the comet in his imagination he then takes the reader into the cold depths of the solar system before speeding inwards to visit the planets.

The Gas Giants

"Neptune readily drags us out of our course. Here we may discover objects that have escaped the keen eye of astronomers. No astronomer has ever detected more than one satellite; but we may well suppose that this arises, not from their non-existence, but from their invisibility at such a distance. There are probably crowds of moons studding the Neptunian skies, and giving cheering light when the tiny sun has set, the sun being only a thousandth part as large as it appears from our globe. It is not improbable that Neptune has rings like Saturn."

The rings of Neptune were not discovered until 1984 by three astronomers using the La Silla Observatory in Chile. The Voyager 2 spacecraft blazed its way past Neptune in 1989. The count of Neptunian moons was elevated from one in Leitch's time to two by Gerard Kuiper in 1949, three in 1981, and eight by the time of Voyager. Today the count rests at fourteen.

A similarly insightful guess appears when Leitch arrived at Uranus.

"Here we find numerous satellites. Sir William Herschel discovered six, but only four have been detected by others. It is, however, highly probable that the number is greater even than that assigned by Herschel."

Today the Uranian moon count sits at 27.

When Leitch arrived at Saturn in his imagination he concluded:

"An opportunity is now afforded of inspecting the mystery of the rings. You will probably discover many more rings, or, rather, what appears a single ring will be found to consist of many smaller ones. You can see through the dusky ring, and have an opportunity of detecting its nature. You will find it to be different from vapour or gas, and to consist of meteorites of considerable size, though, at the dis-

tance of the earth, it would appear as if you were looking through a cloud of fine dust. It is probable, also, that you will find the brighter rings to be of a similar nature, though the bodies of which they are composed may be larger and more closely packed together. The rings have, not without reason, been suspected to be rows of satellites, so closely crowded together that they appear to be one solid body. This accounts for the occasional appearance of divisions, and their subsequent obliteration."

Asteroid Colonies

His assumptions about the asteroid belt are in a similar vein. Only 71 asteroids were known in his time but he supposed that there must be thousands. He even postulated the construction of a human habitat inside an asteroid.

"The very globe itself might be tunnelled and split up, so that contending parties might have little worlds of their own to live in. The imagination can thus easily revel in the wildest fancies, if we exchange the normal conditions of life for extreme physical suppositions."

Chapter 9

Light and Time

After touring the solar system Leitch proposed that the reader should then embark on a journey to the stars.

"We have still an expedition before us, which may be compared to the crossing of the Atlantic, or a voyage to China. We have not yet really left home, and now that we propose going abroad, what vehicle shall we take to aid us in our flight to other systems? The comet is all too slow for our purpose. We must have something still more subtle and swift. The only physical agency that can serve our purpose is a ray of light...Let us suppose, then, that, with the ethereal vehicle of light we are to start upon a journey far beyond the solar system, where shall be our first resting-place? Alpha Centauri is the nearest of the stars whose distance has been well determined; but with all the spiritual swiftness of light, we can reach it only in three years and a quarter."

This "vehicle of the imagination" seems to have resonated down the decades and is highly reminiscent of the *Cosmos* television series created by Cornell astronomer Carl Sagan in the 1980s. It is certainly crystal clear that Leitch understood the formidable barrier presented by stellar distances, but his comprehension of light-speed is almost as remarkable as his other insights.

"From the simple law, that light requires time to travel from one point to another, it follows, that we see everything in the past. In the case of very distant objects, this leads to startling results. For every event in the past history of the world, there is a corresponding point in space, and if we were situated on a star at that point, we would, on looking down upon the earth, see the corresponding event transacted.

"For example, if we took up our position in a star, to which light would take six thousand years to travel from this globe, we would witness the scenes of paradise, and the roll of the world's history would unfold itself to our eyes. If the course of events appeared too slow, we could hasten it, in any degree, by gliding swiftly towards the earth, along the course of the rays. If we could accomplish the

journey in an hour, the history of six thousand years would be condensed into that period.

"The schoolmen defined eternity as punctum stans, *and the propagation of light gives a startling illustration of their meaning. We can arrest the flow of time by continued motion. Suppose our world is the illuminated dial of a clock, that the hand is at twelve o'clock, and that the machinery is faithfully doing its duty; we have only to take up our position in a star that moves from the earth as rapidly as the rays from the dial, in order to arrest the hand for ever at that hour. To one who is stationary, the hand makes its ordinary revolution; but one who moves away with the rapidity of light, sees it perfectly fixed. Nay, it is possible to turn back the hand, as in the case of the dial of Ahaz.*

"In a star moving away from the earth more rapidly than the light, a person would see the hands gradually move in the reverse order, from twelve to eleven o'clock, and so on. By moving in the direction opposite to that of the light, centuries might be concentrated into hours, and hours into seconds. Had we unlimited powers of locomotion, we would not be under the necessity of reading unintelligible and prosaic accounts of campaigns and battles in the past history of our country; it would only be necessary to wing our way to some star where the light from the seat of war is just arriving, and leisurely watch the actual progress of events."

It cannot be stated with any certainty that these are original thoughts by Leitch but they serve to illustrate his grasp of complex concepts. There was no Einstein or *Relativity* when this was written. Nor was there any concept of the speed of light being an absolute, thus preventing any possibility of outpacing it. Relative speed had been discussed quite coherently by Nicholas of Cusa in 1440, but this is more than that. It is a lucid description of the special relationship of light with time, and his idea of perceiving frozen time, due to a relative state, is only one step removed from Einstein's "thought experiment" of a frozen light wave.

The Size of the Universe and the Great Debate

In one other comment Leitch explained:

"But from our (observation) position we find that the Milky Way,

with its millions of stars is not the only luminous disc. The whole heavens are studded over with similar patches of light or nebulae which, on closer inspection, are found to be firmaments, consisting, like the Galaxy, of innumerable stars. They may appear as single, hazy stars, but they are the combined light of countless hosts. These groups are separated by gulfs which it would require millions of years for a ray of light to traverse."

Again we seem to see Leitch pushing himself out of his own time. The only method for detecting stellar distances at this time was Bessel's parallax system. William Huggins would not use the Doppler Effect to detect relative motions in stars until four years after Leitch's death. Henrietta Leavitt would not discover the special properties of the Cepheid variable until 1912. Both discoveries would help astronomers to get a grip on the size of the visible universe.

Furthermore, on April 26th 1920 an important debate took place at the US National Academy of Sciences. In fact this was such an important debate it became known as "The Great Debate". The two contestants were astronomers Harlow Shapley and Heber Curtis and they had literally come to spar over the scale of the universe. Their best estimates were derived from the latest technology and science.

Curtis had researched the frequency of novae in the Andromeda "nebula" and he concluded that there were far too many exploding stars for such a small area if it were inside our galaxy. In fact the numbers seemed comparable to those observed within the rest of the entire Milky Way. He had to conclude that this meant the Andromeda "nebula" was in fact another galaxy. If that were the case then the other spiral nebulae could be anything from 500,000 to 100,000,000 light years distant. This sort of debate had been ongoing for decades, long after Leitch (and Nichol) had already accepted that millions of light years separated the galaxies.

Leitch concluded his extraordinary tour-de-force with this.

"When we step from planet to sun, from sun to system, and from system to firmament, we are ascending the rounds of the ladder that leads up to the Infinite; and this is the great end of the book of God in the heavens. But a hard-featured philosophy comes, and tells us that we cannot know the Infinite, that the notion we form is merely a synthesis of finites, that no number of finites can ever make an

infinite; and that this arises from the very limits of thought. This is true, if it means merely that we cannot construe to our minds the image of an infinite ladder, by indefinitely increasing the rounds of it; but surely we can know a thing, though we cannot draw a definite picture of it to the eye or the imagination."

Despite, or perhaps because, he was a devout and pious man Leitch was clearly not going to be intimidated by the Cosmos. He even dipped his toes into the highly controversial anthropic principle when he stated, *"What is space? Is it an objective reality, or a subjective condition of thought? We cannot enter on this mare magnum of controversy…"*

"A Journey Through Space" was first published by Strahan in "Good Words" in September of 1861. It was evidently presented to Macleod during Leitch's trip to Scotland that summer.

A review on the 12th September stated that this issue of Good Words was, *"the best sixpenny worth of literature ever placed within the reach of the British public."*[1]

Notes:

1.   Stirling Observer Sept 12 1861

Chapter 10

Back to Canada - 1861

In Mid October of 1861 the Presbytery in Scotland were anticipating Leitch's official resignation from his Ministry in Monimail. A debate spilled out into the pages of the newspapers as to whether the Presbytery should accept his resignation or fight it. When the time came his resignation letter was printed in full in the newspaper.

"Rev. and Dear Sir, Having received the appointment of the Principalship of the University of Queen's College , Kingston, Canada, I hereby resign and demit the office of minister of the parish of Monimail, and beg the rev. the Presbytery of Cupar to receive and sustain this my resignation and demission; and, having done so, to declare the vacancy occasioned thereby, that the proper steps may be taken for providing the said parish with another minister.

"In tendering my resignation, I cannot but express a warm acknowledgement of the friendly and cordial intercourse I have enjoyed among the members of the Presbytery during the last eighteen years. Being now called, in the providence of God, to labour in a distant part of the world, I shall bear with me the pleasant recollection of being so long associated with brethren whom I much loved and esteemed.

"May the blessing of the Most high rest upon you and upon your families and flocks. May you have many seals of your ministry; and, after an active life spent in the service of our common Master, may you have an abundant entrance into the heavenly rest. – I am, Rev. and Dear Sir, yours very truly, William Leitch."

A debate ensued and at least one party referred to Leitch as the *"oil which made the machinery of the court run smooth."* Another alluded to his *"articles in 'Good Words' which were of such a nature as to raise him greatly in the estimation of the public."* And yet another said that Leitch intended to establish the first astronomical observatory in Canada (which was of course incorrect.) Leitch himself still worked on his daily tasks and delivered a sermon in Flisk about *"the work of revival in Canada and New York as witnessed by himself. The church was crowded; and the audience were thrilled by the eloquent illustrations of the Reverend Doctor."*

On November 9th the Presbytery accepted his resignation and that night a dinner was held in his honour at the Royal Hotel in Cupar. When called upon to speak, Leitch expressed his high opinion of Canada before stating, *"I shall have ample opportunity for carrying out the chief study of my life - the connection of science with religion."* Ministers from a multitude of denominations made flattering speeches and voiced their regret that he was leaving. One stated, *"In the West Country, where he resided, he was regarded as one of the best artistic and scientific writers to the periodical press that they had at the present day."* [1]

Unbeknownst to Leitch, just a few hours before his party in Cupar, on the other side of the Atlantic Ocean, the USS San Jacinto had just opened fire on the British mail packet RMS Trent off the coast of the Bahamas. The Americans boarded the British vessel and arrested two Confederate emmissaries who had managed to break through the Union blockade and were on their way to England to drum up support for their cause. The incident almost triggered all-out war between Britain and the Union.

The San Jacinto waylays the Trent (Nov. 1861)

Almost immediately tensions were raised on both sides of the Atlantic. As the rhetoric exploded into the public eye Leitch returned to Canada. By December 13th 1861 he was back at his desk and writing to offer his assistance to Dr Egerton Ryerson in Toronto who was leading the charge regarding the problems with the University endowment. *"Now that Canada is the Land of my adoption, I am ready to lend my aid in any way calculated to promote the elevation and efficiency, of our Educational Institutions."* [2]

It appears that he must have left another essay in the hands of his publisher in Scotland entitled *"Uses of the Moon"* which they printed in *Good Words* in February 1862. In this essay he investigated the effects that the moon has had on past and present society. Providing light in places without gaslights, operating the tides to flush rivers and keep them pure, providing a navigation beacon for determining longitude, and as an inspiration for lovers and poets. It is a simple and eloquent essay further proving the renaissance mind behind its composition. Evidently his reputation as a polymath also led to him contributing to the noted multi-volume series *The Imperial Dictionary of Universal Biography* published by Mackenzie of London and Glasgow. Once again he was in very select company. Other contributors to the series included Macleod, John Pringle Nichol, David Brewster and Benjamin Disraeli, who just a few years later would become Britain's Prime Minister.

The University Question Continues

William Leitch's life would never be easy. Having injured himself at a young age; married, only to have his wife and two sons die young; and then leaving his remaining two children in Scotland while he tried to make a difference in Canada; things were not going to get any easier for him.[3]

The series of political battles underway in the Canadian educational system never went away for Leitch. He proved that he was not only a brilliant advocate for science but he was also religiously enlightened. He made passionate speeches protecting the rights of people of all religions to be free to worship as they pleased. His actions disproved the slurs made against him by certain people determined to maintain their stranglehold on the endowment. He tried to make peace with the powerful administrators of the Kings College and the University of Toronto in his role as a member of the Senate of that seat of learning.[4]

Because of the events aboard the RMS Trent the British government deployed ships and men to defend Canada. By April 24th Leitch was briefly forced to close Queen's University due to the possibility of a war with the United States. In an address he alluded to the threat *"which had roused the patriotic ardour of the colony, and advocated the continuance of drill as part of the college exercises."*

All through the tense spring of 1862 he sat on the senate board of the University of Toronto, voting on the day-to-day business of the institution as well as acting in a similar role at Queens. On June 7th 1862 he made a passionate speech at the convocation of students in defense of the university; disproving all of the spurious accusations made against him by those who branded him as a man who just wanted to dismantle the university in favour of denominational religious schools. *The Globe* newspaper would report of his performance.

"Rev. Dr. Leitch also replied. In the course of a remarkably good speech, the reverend gentleman said that one of the things which most surprised him when he first came to Canada, was the antagonism existing between the academical institutions of the country. He could not see, and did not now see why they should not form one great brotherhood, instead of being ranged in deadly hostility one against the other. (Loud cheers.) He argued with his colleagues of his own (Queen's) college on this subject. He said, "Let me go up to Toronto and judge for myself, perhaps there are good fellows in Toronto." They shook their heads very doubtfully, but he carried out his intentions, came to this city and exercised his functions as a member of the Synod. He found among the gentlemen here, the utmost desire to unite with the various institutions throughout this country, in order to form one great institution. This was the basis of the negotiations of last winter, and the feeling existed not only here but throughout Canada. The prime consideration was not to establish a scheme of spoliation, but on the contrary, to make the University of the greatest possible efficiency. He was the last man in any way to cripple this institution. He would be a traitor to his country, a traitor to literature and science, if he did anything of the sort. He

believed this institution did its work with the utmost efficiency and with the greatest fidelity. (Cheers.)" [5]

Just a few days after what seemed like a good outcome to this intractable problem Leitch left for Scotland again.[6] He would remain there from June until December of 1862 visiting his children who were now being educated at Edinburgh. By the time he left, Abraham Lincoln, in response to a carefully crafted letter from Prince Albert, had chosen to release the two Confederate diplomats, effectively defusing what could have been a catastrophic situation. Leitch must have felt safe enough to cross the border and once again he travelled through Boston and visited his friend Alvan Clark. This time he found both the city and his friend changed irrevocably by the civil war. Clark had finished the lens for his super-telescope and had already put it to good use, his son had just discovered the small companion star Sirius-B a few months earlier, but he was now devoted almost entirely to making *"small telescopes to be fixed on the barrels of rifles."* Astronomy, optics and ballistics must have consumed their conversation. Both the Hale rocket and the Minié rifle would be utilised to devastating effect during the American civil war. [7]

Troops destined for Canada at Woolwich, 1862

Back to Scotland...again - 1862

During his crossing to Scotland Leitch was astonished to find Confederate soldiers on his ship. They had somehow once again breached the blockade and boarded in Halifax, but there were also some very upset Union soldiers aboard. Leitch feared that trouble might ensue, *"but a lady of Boston acted as mediator between the parties and, thanks to her persuasive powers, she charmed down the rising storm, and succeeded in keeping the peace."*

Shortly after returning to Scotland, Macleod published Leitch's essay *"At Night in an Observatory"* in the September 1862 issue. It included a vivid description of the Horselet Hill observatory and the arduous work undertaken there; clearly written by a man who had lived through similar experiences. He also described the arrival of the first automated electric "eye" used by astronomers at Kew, the photoheliograph, invented by Warren de la Rue, which was the first true astronomical instrument specifically designed to take photographs. De La Rue had exhibited images from his device at the 1862 International Exhibition, which was in full swing while Leitch was in Britain, which again suggests that Leitch probably attended the exhibition where Hale's rockets were on display. Leitch warned his readers not to think of the photoheliograph as "a wild conception of Frankenstein" but a useful and promising new tool. This brief comment shows that although Leitch predates Jules Verne, he had evidently read Mary Shelley's masterpiece, a book many consider to be the first true modern science fiction story.[8] It's not entirely surprising that Leitch was familiar with Frankenstein. The author of the famous gothic horror novel, Mary (Godwin) Shelley, developed her taste for adventure as a teenager while living in Scotland with a family named Baxter. The Baxters knew Shelley's father - the famous writer William Godwin, and they shared a similar world-view; one in which women were respected. During her long sojourn in Scotland Mary Shelley became life-long friends with Isabel Baxter, one of the free-spirited daughters of the house. Shelley considered the Baxters to be her foster family.

By the time Leitch arrived in Monimail Isabel Baxter was living in the neighbouring village of Balgonie with her husband, daughter and son-in-law Joseph Gordon Stuart. Just two months after being ordained as minister of Monimail Leitch was asked to join a group of lecturers who were speaking out on the subject of voting rights. Stuart was the president of the group. Leitch could hardly have met a more kindred spirit. Stuart was also the Chair of the Fifeshire Association of Independent Churches, the Chair of the Parochial School Board and later a familiar face on the local lecture circuit talking on subjects ranging from science to history.[9]

The so-called "Common Suffrage" movement had been given a substantial boost in 1843 and evidently Leitch and Stuart both felt that it was important to support the right of all people to vote; including women. Predictably Stuart had been attracted to a family that encouraged the enfranchisement of women and must have known all about his mother-in-law's friendship with one of the most famous writers in the world. A woman whose mother Mary Wollstonecraft was one of the true pioneers of women's rights.

In 1846 Isabel Baxter's husband died. Shelley had tried to help her by pressing the government into providing a pension for the family. Right up until her death in 1851 Mary Shelley remained devoted to her friend. One of her last wishes was that her son arrange for a £50 stipend from her estate to go to Isabel. In 1877 Isabel's grandson married William Leitch's daughter. Isabel's great-grand children were William Leitch's grandchildren. In 2016 the story of Isabel and Mary was filmed for a motion picture called *Mary Shelley*.

A Winter in Canada

During the spring of 1862 Leitch had written another eloquent essay named "*A Winter in Canada*" in which he beautifully described the Canadian environment, while touching upon such issues as the safety of the trains and simultaneously showing his understanding of $E = \frac{1}{2}mv^2$. It first appeared in the December 1862 issue of *Good Words* and would subsequently appear in a Macleod anthology in 1871. [10]

"One train may run into another, or get off the track, but the injury is comparatively slight when the speed is only fifteen or twenty miles an hour. When the speed is increased, the destructive power increases in a much higher ratio. When it is doubled, the destructive power is increased fourfold; or, in other words, the destructive power increases as the square of the velocity."

During one difficult journey when the train was stopped by bad snow he made observations of the powder piling up on the windows.

"The process of formation was all visible before you by looking minutely into the structure. With the aid of a glass you could see the minute particles rolling over and taking their allotted place, so as to fit into the general pattern. Some of the forms were not unlike the spiral nebulae which Lord Rosse's telescope has revealed to us; and the great laws which pervade the universe can as well be illustrated by the play of atoms as by the revolution of worlds."

It is fair to assume that during his long sojourn in Scotland in the summer of 1862 he wrote and compiled five more essays about space. They were: The Chemistry of the Sun; The Structure of Saturn's Rings; Stellar Grouping; The Stability of the Solar System; The Eternity of Matter; and he seems to have finally decided to publish his thoughts on The Plurality of Worlds. None of these essays appeared in *"Good Words"* they became the final chapters for his forthcoming book.

God's Glory in the Heavens - 1862

Leitch's book would finally be published in a run of a thousand copies on November 22 1862.[11] It would be sold in every major English-speaking country. Editions were available in Canada, the USA, Australia and even India. Reviews were generally favourable, but the title had been reduced to simply *"God's Glory in the Heavens"* with no reference to astronomy, thus setting the stage for it to effectively vanish in its complete form. The only clue that the book contained scientific insights was an embossed image of Herschel's telescope on the cover. This may explain why it has been overlooked for so many years by secular scientific researchers. The book is almost certainly filed under theology in many libraries.[12]

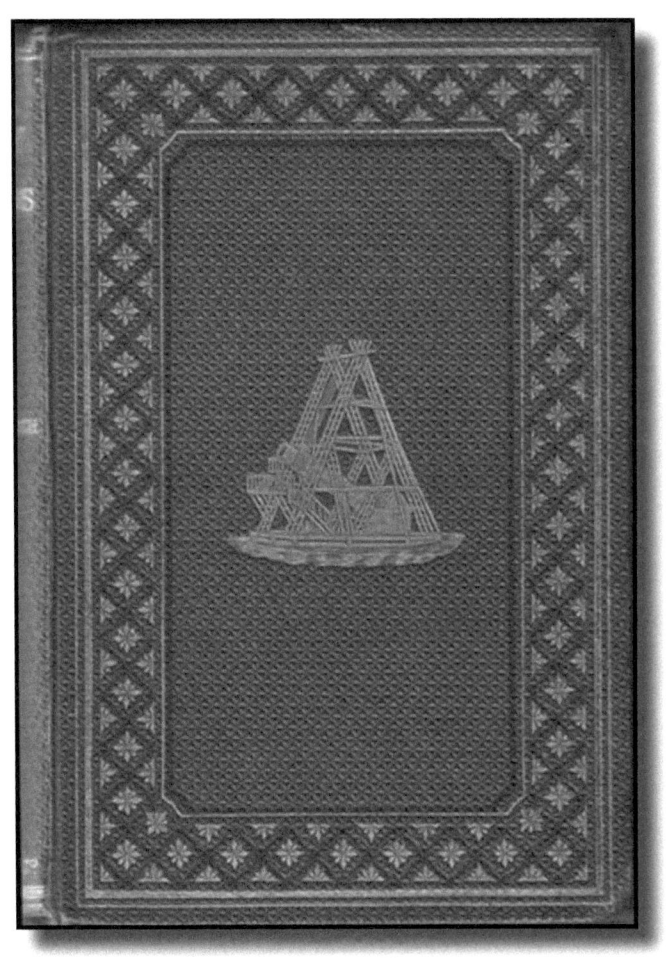

The 1st Edition of
God's Glory in the Heavens (1862)

Notes:

1. Fifeshire Journal, Nov. 7, 1861; Glasgow Herald, Dec. 4, 1781

2. Documentary History of Education in Upper Canada by J. George Hodgins, Vol 16 and 17, Cameron, Toronto, 1907

3. Ancestry.com

4. The Journal of Education for Upper Canada, Vol. 17, 1864

5. Globe, Toronto, Jun. 7, 1862

6. Glasgow Herald, Nov. 10, 1862

7. Clark held patents on rifle loaders and was also an expert on ballistics. He was acquainted with the world's leading astronomers including John Herschel and Lord Rosse. A Clark telescope discovered the moons of Mars.

8. At Night in an Observatory by William Leitch, Good Words, Strahan, London, Sep. 1862. (Strahan had relocated from Edinburgh to London in June 1862.)

9. Fife Herald, 23 Mar 1843, 20 April 1843, 23 Nov 1843, 14 Dec 1843, 3 Feb 1848, 20 Dec 1849

10. A Winter in Canada by William Leitch, Good Words, Strahan, London, Dec. 1862; Peeps At Foreign Countries, Strahan, London, 1871

11. The Aetheneum, Nov. 22, 1862

12. God's Glory in the Heavens by William Leitch, Strahan, London, 1862

Chapter 11

Back to Canada to Stay - 1863

In December of 1862 Leitch returned to Canada aboard the Cunard liner *Europa* travelling from Liverpool to Boston.[1] He was accompanied by Reverend J.C. Murray and the two arrived at the Kingston docks aboard the lake steamer *"Pierrepont"*. They were greeted on the wharf by the divinity students and by arts students in full academical attire.

The Great Lakes Steamer Pierrepont (ca. 1862)

That same month a newspaper story announced that he was being considered for the prestigious position of the Chair of Divinity at Glasgow University.[2] It must have been a post that he would have given serious consideration, but he was not scheduled to return to Scotland until the following summer. His two children still remained there and he had never purchased a home in Kingston, preferring to rent accommodation and operate his business from an office at Queen's. Despite never really establishing roots in Canada Leitch continued to press as hard as he could to pull Queen's out of its political travails and establish some sort of order to the flow of students, examiners, honours and money in the Canadian school system.

Between March and May of 1863 the battle over the endowments continued to rage. Despite his efforts of the previous year there was still an entrenched group who preferred the status quo and Leitch's gentle articulate nature made him an easy target.[3] On April 17th 1863 Leitch wrote to Ryerson to say that he had just concluded a meeting with John A. Macdonald and Alexander Campbell about the university question and he had been advised to back down. Despite the settlement between the USA and Great Britain, things were still extremely tense for Macdonald. Canada's leading politician thought it impolitic for him to take the time to fight the school cause because, *"We are on the eve of an election contest which may determine the future of Canada—and whether it will be a limited Constitutional Monarchy or a Yankee democracy."*[4] A few weeks later while Leitch was in Quebec the latest in a series of votes brought an end to Ryerson and Leitch's cause.[5] Without Macdonald's support the fight was lost. However, more than a generation later many of their proposals would be adopted.

It isn't known for certain if Leitch returned to Scotland again in the summer of 1863 but all indications suggest that he probably did not. His book had sold out and in September of that year a second printing of another thousand copies hit the streets in Scotland and England.[6]

In November Leitch pushed for further reforms when he arranged to put the power of hiring and firing into the hands of the school's Board of Trustees (over which he presided) so that no professor could slander the school's reputation with impunity. He followed this with the approval of a publication entitled *"Defence Of The Plan Of University Reform"* which outlined all of the logical reasons why Canada's institutes of further education should share the bounty of the endowment and how this would be to the benefit of all the denominations, including atheists.[7] In the autumn he travelled to synods in New Brunswick and Nova Scotia and then spent Christmas with friends in Montreal. It was probably during this trip that noted pioneer and royal photographer William Notman took a photograph of Leitch for inclusion in his book, which was published in 1865.[8] It was also when he probably first exhibited signs of his impending illness.

Since arriving in Canada he had undertaken to wage the same battle he had fought back in Fife. To try and make sure that public money went to the places to which it was assigned. He had no idea that the battle at Queen's in Kingston, which had been raging for years, would become such an intractable problem. He brought all of his intellectual capital to bear while still performing his duties as Principal, teacher, Board Member, Senator at the University of Toronto and head of the Synod. He managed to establish a more permanent astronomical observatory at Queens. He associated closely with his fellow Scots, Macdonald and Campbell and worked diligently to establish a fair system of education that respected all of the denominations of the church, but his battle to try and make an equitable education system in Canada was taking its toll.

Leitch was also never far from the political strife which had begun in Scotland. He was either too pious for some or not pious enough for others. On one occasion news leaked out that he had been celebrating the Prince of Wales' birthday on a ship during the Sabbath. This apparent breach of etiquette made it into the newspapers in Scotland before someone forced a retraction to the story. It wasn't true. Leitch was a man of dedication, both to science and to his faith. He saw no contradictions between the two. He advocated the position that if science seemed to contradict the Bible, don't denounce the science, just don't make the inference that if the science is right then the Bible must be wrong. Find a third road and assume that both can still be right. In this regard he was, along with people like William Whewell, one of the last of the true natural philosophers.

The Final Months – Winter 1863 to Spring 1864

"About Christmas last, in the course of his fourth session, Principal Leitch went to Montreal on a visit to some friends; and thence to Ottawa, for the purpose of attending a Bible Society meeting. At the latter place, allured by the magnificent winter scenery, he took a long ramble, and after being over-heated sat down to rest. In the evening of the same day, in the middle of a speech reported to have been unusually effective, he was carried by spasms of the heart." [9]

The Quebec Daily Mercury - Apr 19, 1864

ILLNESS OF PRINCIPAL LEITCH.

The illness of Principal Leitch, of Queen's College, we regret to announce, has assumed a fatal turn, and it is believed by attending physicians that he can survive but a brief period. His dissolution is hourly expected.

A few weeks later, in February 1864, Leitch suffered another heart attack. His contemporaries maintained it was the stress of the university fight which had brought him to ill health. He struggled to return to work and attempted to fulfill his duties, but stepped down as the teacher of Metaphysics and Ethics on April 26th 1864. [10][11]

Leitch would not live to see his children grow up. He died in Kingston Ontario on May 9th 1864 just before his 50th birthday. His role of explaining mankind's role in the universe to the readers of Good Words was immediately taken over by Sir John Herschel, the very same man who had met with Darwin 28 years earlier. Leitch was buried in a new plot at Cataraqui Cemetery in Kingston.

The obituaries spread from Canada to Scotland were universally kind to him. Even the Globe and Mail which had spared no effort to publish anonymously-scribed articles damning his efforts on behalf of Queen's College were willing to grant him a generous departure. His death was attributed to heart disease. When he died his friend and editor Norman Macleod would write, *"I have lost a dear friend in Principal Leitch. Poor dear Boss! I cannot think of the world as henceforth without him—so simple and true, so loyal, so genuine! I have, with very few exceptions, no such friend on earth—none who knew my failings as he did, none to cover them as he did, none to love me in spite of them as he did. Well, he is another portion of my treasure in heaven!"* [12]

In the Fife Herald, which had also once worked hard against his efforts on behalf of the Parish schools, the comments were universally kind, mentioning his time teaching astronomy and science at Glasgow and how his strength lay in "scientific theology." The Dundee papers reported, *"...his habit of calm philosophic thought and his power as a deductive reasoner, with the abundant stores of illustrative material at his command in natural science and psychology, (he) would have done honour to the discrimination of his patrons and enhanced the reputation of his alma mater."*[13]

The Paisley papers reported from Kingston:

"After finishing his preparatory studies for the Church of Scotland he did not immediately enter on the practical work of his profession, but remained for some years in connection with Glasgow Observatory under the late Professor Nichol. The science to which he remained most fondly attached was that of astronomy, and from his thorough familiarity with the practical working of an Observatory, from the enthusiasm with which he studied every improvement in astronomical instruments, and hailed every fresh discovery to which it led, as well as from his general scientific attainments, it was probable that, had he not left Scotland, he would have been appointed to the chair of his teacher, the late Professor Nichol, in the University of Glasgow. De Quincey, in a noble article on Lord Rosse's telescope, speaks of his friend Professor Nichol, as having contributed more than any living man to keep general English readers, who have not time for the scientific investigations of astronomers, acquainted with the latest and profoundest results these investigations are leading; and during the two years which have passed since the Professor's death, it would have been difficult to point to a man for whom the same distinction could have been so justly claimed as the late Principal of our University." [14]

Notes:

1. Glasgow Herald, Jan. 28, 1863

2. Globe, Toronto, Dec. 8, 1862

3. Ibid. May 14, 1863

4. Egerton Ryerson His Life and Letters, C.B. Sissons, Clarke Irwin and Co, 1947

5. Documentary History of Education in Upper Canada by J. George Hodgins, Vol 18, Cameron, Toronto, 1907

6. The Bookseller, London, Sept. 30, 1863

7. Globe, Nov. 13, 1863

8. Portraits of British North Americans by William Notman, Montreal, 1865

9. Fife Herald May 26 1864

10. Globe. Feb. 13, 1864

11. Documentary History of Education in Upper Canada by J. George Hodgins, Vol 18, Cameron, Toronto, 1907

12. Memoir of Norman Macleod by Donald Macleod, Worthington, N.Y., 1876

13. Dundee Perth and Cupar Advertiser May 27 1864

14. Kingston Daily News May 11 1864

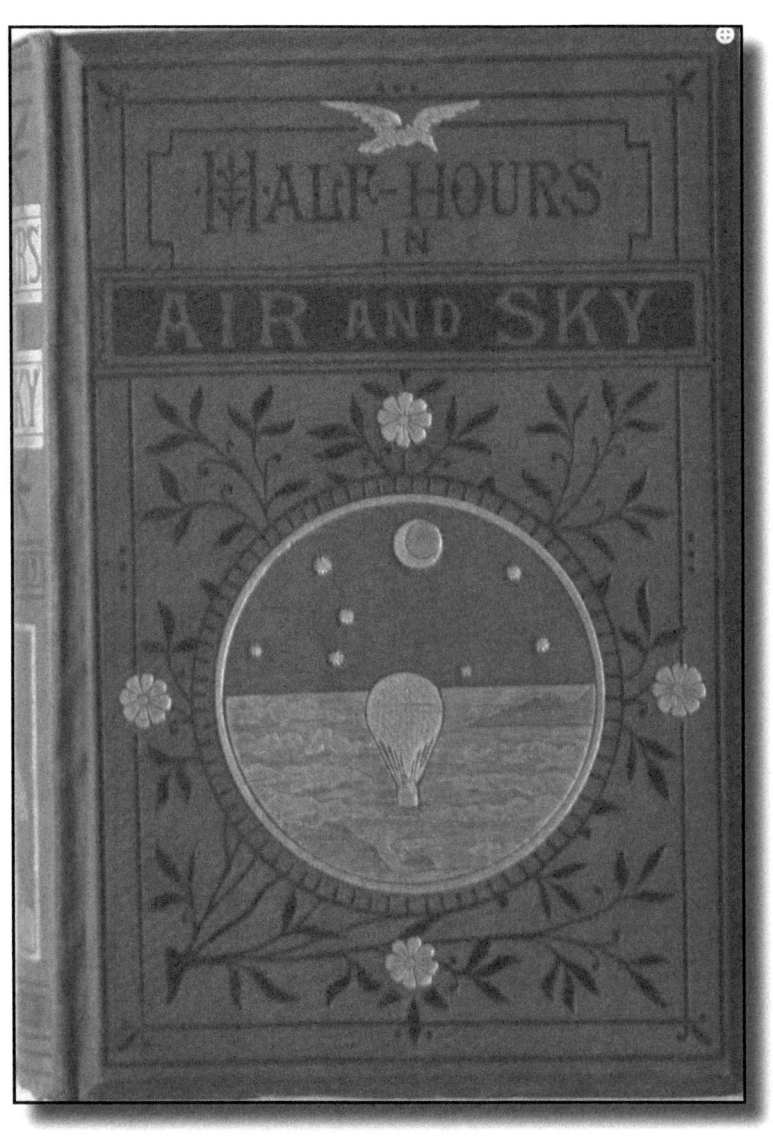

An 1882 Edition of the anthology which included Leitch's "A Journey in Space".

Engraved by Samuel Sartain Phil.ª

W Leitch

Chapter 12

Conclusions

Leitch's book would see its third edition, first in New York in 1866, and the following year in England. The fourth edition appeared in 1875. It was still in print until at least 1877 and as has been previously noted, it would then be plundered and anthologised for at least another three decades beyond that; evidently always without attribution.

This book would be his one great work in a short life of only 49 years. The new chapters he added to the book in 1862 included his efforts to address the age and nature of the universe and his long developing theories on the possibility of alien life; both subjects guaranteed to rock the halls of the literalist interpreters of Christian scripture.

The Eternity of Matter

His essay *The Eternity of Matter* was an attempt to unravel the metaphysics and conflicting philosophies in vogue at that time. After a short assault on those foolish enough to try and create definitions for infinity, Leitch in his usual way touched on the heart of the matter. His peers were battling over the evidence for intelligent design in the universe, a battle which in some quarters continues to this day. After some had argued that the non-eternity of matter was a purely scriptural truth, beyond the reach of human reason, Leitch postulated an insight that, despite being grounded in his belief in a Creator, could almost be taken as a description of the Big Bang, *"If we admit the argument of design...may we not legitimately push the argument somewhat further, and hold that matter itself must have had a beginning? For example, the solar system manifests design, and had, therefore, a beginning; but the matter out of which the system was formed must have been wisely adapted for such a cosmical combination; and are we not entitled to infer that this chaotic matter had a beginning also, or that matter is not eternal?"*

Again we can see him ploughing headlong into the biggest question of all; a question still evading a clear answer 150 years later.

The Logic of the Plurality of Worlds

Finally, when addressing the notion of alien intelligence Leitch once again challenged many of the evangelical community by shredding all of their assumptions derived from scripture. After lecturing on the subject for a decade we can assume that this chapter of his book distilled down all of his conclusions. Despite believing in a Christian universe Leitch was not willing to fall into the trap of assuming that a version of the Christian Saviour was dropping in on every planet on a mercy mission. Nor was he willing to accept that the universe was made solely for humans. He proposed that it might just as easily be possible that *only* the Earth and mankind had committed mortal sins and that we might well be the black sheep of the galactic community. [1] He argued in favour of alien intelligences in a multitude of forms because the Bible not only didn't preclude them, it described them - as *Angels*. In his final analysis Leitch used his own faith to rationalise why we seemed to be alone in the Universe. He suggested that perhaps only mankind had fallen into God's bad books and we were consequently ostracized by the other galactic intelligences until we had redeemed ourselves. It was a poetic and typically logical explanation, a third path, from a man restricted on two fronts, by faith and science.

Leitch's old schoolmate William Thomson would not be so restricted by scripture; leaving behind the old philosophies and embracing the world as a true scientist. He went on to become President of the British Association. Many years later he would be one of the first to postulate that life actually came to earth from space.

It was a theory that still prevails in some circles to this day. Here it is in Thomson's own words, *"How did life originate upon the Earth? Tracing the physical history of the Earth backwards, we are brought to a red-hot melted globe on which no life could exist. Hence, when the earth was first fit for life, there was no living thing on it. There were rocks, water, air all around, warmed and illuminated by a brilliant sun, ready to become a garden. Did grass and trees; and flowers spring into existence, in all the fullness of a ripe beauty, by a fiat of Creative power? Or did vegetation, growing up from seed sown, spread and multiply over the whole earth? Science is bound by the everlasting law of honour, to face fearlessly every problem which can fairly be presented to it. If a probable solution, consistent with the ordinary course of nature can be found, we must not invoke an abnormal act of Creative power.*

"When a lava stream flows down the side of Vesuvius or Etna it quickly cools and becomes solid; and after a few weeks or years it teems with vegetable and animal life, which – for it – originated by the transport of seed and ova, and by the migration of living creatures. When a volcanic island springs up from the sea, and after a few years is found clothed with vegetation, we do not hesitate to assume that seed has been wafted to it through the air, or floated to it on natural rafts.

"Is it not possible, and, if possible, is it not probable, that the beginning of vegetable life on earth is to be similarly explained? Every year thousands, probably millions, of fragments of solid matter fall on the earth. Whence come these fragments? What is the previous history of any one of them? Was it created in the beginning of time an amorphous mass? This idea is so unacceptable that, tacitly or explicitly, all men discard it. It is often assumed that all, and it is certain that some, meteoric stones are fragments which had been broken off from greater masses and launched into free space. It is as sure that collisions must occur between great masses moving through space, as it is that ships steered without intelligence directed to prevent collision could not cross and recross the Atlantic for thousands of years with immunity from collisions. When two great masses come into collision in space, it is certain that a large part of each is melted; but it also seems quite certain that, in many cases, a large quantity of debris must be shot forth in all directions, much of which may have experienced no greater violence than individual pieces of rocks experience in a landslip or in blasting by gunpowder. Should the time come when this earth comes into collision with another body, comparable in dimensions to itself, be when it is still clothed as at present with vegetation, many great and small fragments, carrying seed and living plants and animals would, undoubtedly, be scattered throughout space. Hence, and because we all confidently believe that there are at present, and have been from time immemorial, many worlds of life besides our own, we must regard it as probable in the highest degree that there are countless seed-bearing meteoric stones moving about through space. If at present

no life existed upon this earth, one such stone falling upon it might, by what we blindly call natural causes, lead to its becoming covered with vegetation.

"I am fully conscious that many scientific objections can be urged against this hypothesis; but I believe them to be all answerable, - the theory that life originated on this earth through moss-grown fragments from the ruins of another world may seem wild and visionary; all I maintain is, that it is not unscientific."

A Night in an Observatory

In Leitch's 1862 essay "A Night in an Observatory", he reminisced about the long cold nights spent watching the sky at Nichol's Glasgow observatory and almost lamented the loss of his naivety regarding the Cosmos. *"What should be the most marked moral feature of the astronomer's character? One would think that, as it is the business of their lives to look up to this inverted hand of God, they would be habitually impressed with His glorious presence... (but) the sailor on the masthead of a ship-of-war is in a better position for forming a right judgment of the battle-field and the glory of the victory, than the man who is in the thick of the fight."*[2]

Legacy

Long after Leitch's death, John Nichol, the son of Leitch's astronomy teacher, recalled, *"Mr. Leitch, had distinguished himself by the display of unusual mathematical talent, and attracted my father's attention in the astronomy class. He did all his work with an admirable perseverance, and a degree of slowness almost exasperating. Some indolent people alternate idleness with an amount of activity at times; others, like large bumble bees, never idle, nor ever active, drone through a good day's work every twenty-four hours. Of this latter order Leitch was. I see him like a shadow in the past, jotting down with slumberous accuracy the records of the hygrometer, no morning was wet enough to damp his ardour; or noting the transit of the stars, no night was cold enough to make him shiver; I hear him, as we are discussing metaphysics together over the fireside of Monimail... I fancy him delivering his address as Principal of King's College, Toronto, (sic) and the images are of the same substantial, excellent man of this world and the next."* [3]

So what can be deduced from the life of this likeable Scottish-Canadian minister? The context provided in the first few pages of this book serves to demonstrate how Leitch was chronologically far ahead of Tsiolkovsky and Goddard, but it still doesn't persuasively position him in a place and time. The fact that Charles Dickens, a hero of Jules Verne's, had just written *Great Expectations* may have some resonance for the reader; or that his old schoolmate had yet to invent the Kelvin temperature scale. Or perhaps more salient to the discussion would be a quick survey of the state of the art of rocketry.

When Leitch wrote of his space rocket the British engineer William Hale had only just received his patent for his rotating artillery rocket. The earliest metal rockets, powered by black powder and steered by a stick, had appeared in battle during the Napoleonic wars. They were wild and unruly and as dangerous to the man launching them as to his enemies. Hale's primitive lump of metal and gun powder, about a foot long, was the best kind of rocket available when Leitch proposed riding a rocket into space.[4]

It is perhaps also worth noticing that after Leitch returned to Canada for the last time, his publisher in Scotland felt that it was advisable to continue the monthly astronomy essays. Leitch was replaced by the three time president of the Royal Astronomical Society, Sir John Herschel who began the task of succeeding Leitch by writing a series on comets.

A review of Leitch's book appeared in *The Reader* which said, *"We cannot conclude our notice of Dr Leitch's book without dwelling upon the admirable manner in which the astronomical facts contained in it are blended with practical observations and the highest and most ennobling sentiments. It is thus that books on popular science should ever be written."* The Baptist Review said, *"His papers will never be thrown aside with disgust, but are certain to win favour wherever they circulate."*

In his home parish of Monimail in Scotland Leitch's name is inscribed on the monument where his wife and two sons were laid to rest. In his adopted country of Canada the loss of William Leitch was felt so keenly that his colleagues at Queen's College resolved to erect a monument to him. They wrote to his friends in Scotland who agreed to send £100 if Canada would chip in £200. [5] This led to the establishment of a college plot at the Cataraqui cemetery in Kingston where Leitch was laid to rest on October 4th 1864, exactly ninety-three years to the day before Sputnik.[6] The inscription on the obelisk makes note of the fact that two scholarships were set up in his name by his friends in Canada and Scotland, *to commemorate Dr. Leitch's learning, educational ability and Christian worth.* [7]

The two Leitch memorials Kingston (left) and Monimail (above). He is buried at Kingston.

Tsiolkovsky, Goddard and Oberth all became school teachers. All three taught science and all three were inspired by the fantasy of science fiction and the practicality of science fact. Leitch was also a science teacher. In a world essentially bereft of science fiction his inspiration seems to have been his religion, along with the exciting new technologies of the industrial revolution, and the intractable laws of Isaac Newton.

Ironically, of this illustrious group Leitch was the only one who was truly contemporaneous with Jules Verne.

Perhaps less surprising is the fact that both Verne and Leitch shared a common fascination with Brunel's enormous *Great Eastern* super steamship. After his adventure in Scotland Verne would experience another hair raising passage, this time aboard Brunel's massive paddle steamer. He ended up in Niagara Falls Ontario, in 1867, just at the exact time that he was half way between the publication of his two most famous space adventures; but of course Leitch had expired three years earlier.

Twenty seven years after Leitch's death Canada's first Prime Minister, Sir John A. Macdonald, was laid to rest a few hundred feet away from him in the same cemetery. Today it is a National Historic Site. Despite never sharing in the King's endowment Queen's became a prestigious University and one of its most famous alumni would be Elon Musk, owner and founder of SpaceX.

Despite William Leitch's premature departure, the Leitch family was to prosper back in Britain. Leitch's son John and his daughter Moncrieffe were raised in St Andrews where they developed a taste for golf. John Leitch became a surgeon, first aboard Cunard liners to America, and then at Leith hospital. He had seven children who all knew of their famous grandfather who had gone on a mission to Canada to spread the gospel and the teachings of science. Edith Leitch would go on to fame as a champion golfer and would even meet the son of Abraham Lincoln, perhaps the most famous equal rights leader of the 19th century. Another granddaughter, Josephine, went on to become one of the first female lawyers in Scotland, the family firm still operates today.

Robert Todd Lincoln and Edith Leitch, Sept 1923

However, the most celebrated of Leitch's grandchildren was Cecil, who was by far the best golfer in the family, Cecil played a legendary tournament in the beating rain in October 1910 against British Open Champion Harold Hilton. Cecil lost by only two strokes. It was the first tightly contested game of its kind, since Cecil was actually William's *granddaughter* Cecilia. She went on to win more than a dozen championships in England, France and Canada. On one golfing tour she visited Queen's University because she knew of her grandfather's role there.

Her global fame was such that when she was struck by lightning on a course in Ireland it made the headlines of the Ottawa Citizen half a world away. The story of her almost fatal misfortune appeared just *above* a picture of a fresh-faced young man named Charles Lindbergh who was about to attempt a solo crossing of the Atlantic. Two days later Lindbergh would be the most famous man in the world and would use his fame to encourage the efforts of an unknown New England college teacher named Robert Goddard, who wanted to build a rocket to send into space.

The centennial of Leitch's death would even make the newspapers in 1964. On that day Jack McKay, an American with a Scottish background, rode a rocket-propelled X-15 aircraft to the edge of space.[8]

A fitting place to conclude the story of Principal William Leitch would seem to be an extract from *"Invocation To Professor Boss, Who Fell Into The Crater Of Mount Hecla"* by Norman Macleod.

Oceans of endless bliss Shall roll within thy kingdom;
Cataracts of matchless eloquence shall hymn thy praise;
Mountains of mighty song—mightier by far
Than Hecla, where thine ashes lie entombed,
Shall lift their heads beyond the top of space,
And prove thy deathless monuments of fame;
While thou with kingly, bland, benignant smile,
Look'st down upon the earth's terraqueous ball,
And quell'st with thunder Neptune's blustering mood.

Notes:

1. This notion had been postulated by another famous Man of Fife; St Andrews resident, Dr Thomas Chalmers who had served as a trustee alongside Leitch and George Brewster at the Fife National Security Savings Bank. It was then stretched even further by the noted Christian scholar and writer C.S. Lewis, who in his 1958 essay God in Space argued,"There might be different sorts and different degrees of fallenness." Lewis famously debated Arthur C. Clarke to a stalemate on these issues between 1943 and 1954. Clarke was still willing to let Lewis voice his opinion, four years after his death. He included Lewis' God in Space in his 1967 anthology The Coming of the Space Age". Clearly Leitch would not be the first, or the last, to struggle to reconcile the Plurality of Worlds with his faith.

2. A Night in an Observatory, Good Words, Strahan, Edinburgh, 1862

3. Memoir of John Nichol, James Maclehose And Sons, Glasgow, 1896

4. Chamber's Journal, Volume 3, Jan-Jun, 1855

5. Documentary History of Education in Upper Canada by J. George Hodgins, Vol 18, Cameron, Toronto, 1907

6. Cataraqui Cemetery Archives, Kingston Ontario

7. Documentary History of Education in Upper Canada, Warwick and Rutter, 1907

8. Quebec Chronicle Telegraph May 12 1964

Second Edition – Updates

In 1899, a popular book called "Half Hours in Air and Sky" offered the hope that some discovery would be made to allow "a visit to our neighbor, the moon."

"The only machine independent of the atmosphere we can conceive of," said the book by an unidentified author, *"would be one of the principle of the rocket. The rocket rises in the air, not from the resistance offered by the atmosphere on its fiery stream, but from internal reaction."*

It was a remarkably accurate assessment of how men would travel into space, especially since it appeared four years before Russian scientist Konstantin Tsiolkovsky published his paper showing, for the first time, that a rocket would work in the vacuum of space.

Congressman John M. Murphy of New York, speaking in the United States' House of Representatives, on the eve of the first manned trip to the moon, July 15th 1969.

The above quote was found in the Congressional Record for the 91st US Congress. It clearly demonstrates that Leitch's words continued to resonate right up until the day before the first manned voyage to the moon. As has been demonstrated in the preceding pages, *Half Hours in Air and Sky* was published in multiple editions before and after 1899. This would seem to indicate that since this quote references the same date given by Gatland in 1953, Mark Bloom of the New York Daily News who wrote this for Congressman Murphy was almost certainly citing from Gatland.

In 2016 all of the material available to the author about William Leitch was compiled into the first edition of this book. However, at the time of publication there were many clues to suggest that Leitch wrote and published much more material than seems to have survived. The composition of that first edition required some qualified guesses and assumptions.

In January of 2017 a collection of new material was exhumed and scanned by the diligent librarians and archivists at Queens. This additional material revealed anecdotal snapshots of various events in Leitch's life, as well as letters written back and forth from Scotland about his day-to-day business. However, in amongst the mundane missives could be found little pieces of evidence which further supported some of the guesses and suppositions in the first edition of this book. Only one or two minor things required outright correction. Some of the new information addresses Leitch's whereabouts and acquaintances, and some speaks to his diverse expertise. Most of this information is summarized here to further support the author's opinion that Leitch was an eminent man of science, and should not be lumped in with, as some later historians have referred to them, the "gentleman amateurs" who sometimes frequented Queen's College Observatory.

Reading this section separately may possibly confuse the reader as it makes frequent allusions to the main text contained on the previous pages, therefore page references have been included to try and show how this new material fits into the big picture. To add it all into the main text flow would have required an entire re-write of the book.

Perhaps the most gratifying result of this new research was finding many new essays and lectures by Leitch about astronomy, geology, religion, physics, engineering, philosophy, botany and more. In the interests of simplicity it is presented here in chronological order.

In the Spring of 1848 long before Strahan's *Good Words* appeared Leitch seems to have taken a regular role writing for a magazine published by Myles MacPhail named the *Edinburgh Ecclesiastical Review*. At this time Leitch was quietly living with his new bride Euphemia in the parish at Monimail. Finding these contributions to *MacPhail's* was something of a challenge since Leitch wrote them without any attribution (evidently his ongoing curse). The only clues which led to the discovery of this material was an anecdote which appeared in *The Presbyterian* in 1864 about Leitch having written enough material for MacPhail to assemble an anthology.

MacPhails first appeared in February 1846, *"Under the immediate superintendence of the most eminent Clerical and Lay members of the Church of Scotland,"* so to take in the span of time when Leitch was still in Scotland required the perusal of about a dozen volumes (each over 700 pages). Leitch's essays appeared under the title *Theology and Science* and ran from July 1848 until March of 1851. Spread across 85 pages of content they were broken into 10 separate appearances and incorporated 36 different topics. Considering there is only one book and a couple of pamphlets in print with his name attached to them, this represented a wealth of new material to assess.

Of course the first thing to do was to be certain that it was in fact Leitch's work. If the title wasn't a dead giveaway, the content certainly was. It should be qualified that there is no way to be absolutely certain that Leitch wrote all 36 of these essays unassisted. However, it seems clear that his hand was at work in most of them. There were many clues. For example, in the first essay the author used the metaphor of "climbing the mast" for a better view of the scene. "Climbing the mast" had precipitated a lifetime of infirmity for Leitch and would have been an obvious choice of metaphor, but he also used the same metaphor years later when describing his work as an astronomer (see page 136).

In the same essay the author attacked William Whewell's views on cosmology before discussing LaPlace in some detail (see page 42). He took the extremely liberal position (for a priest) that LaPlace could be a genius and still be an atheist. He expressed his skepticism of assertions without proof (an approach Whewell endorsed a few years later, (see page 46), but in Part 3 of the series he condemned the persecution of Galileo by religious authorities, who defying all logic, insisted that factual proof was not enough to override the dictates of scripture. This was all consistent with Leitch's attitude to scientific logic over-riding any conflicts with scripture (see page 86).

The author of these articles quoted from both Forbes and LaPlace; both were in Leitch's library at Queens. In an essay on the safety of trains the author demonstrated a deft use of mathematical calculus to make deductions about the frequency of railway accidents (see page 121). Clearly this was a person with a solid grasp of mathematics. In a long essay attempting to explain the movements of glaciers he made lengthy explanations about heat theory. We know that Leitch wrote award-winning essays on steam power and clearly understood the fundamentals of combustion. In another he vented on the subject of how the Sabbath should be kept for rest and education, a subject Leitch was known to have fought long hard battles over (see page 31).

Of course all of this is still only circumstantial evidence that this was Leitch, until we come to the eleventh essay which is titled *The Eternity of Matter* written in 1849. The reader might recall that there was a chapter with the same name in Leitch's book in 1862 (see page 133). The 1849 essay covers the same exact discussion of how some of Leitch's peers were attempting to define infinity; this was concluded with the *exact* words which he used more than a decade later.

"For example, the solar system manifests design, and had therefore a beginning, but the matter out of which the system was formed must have been wisely adapted for such a cosmical combination, and are we not entitled to infer that this chaotic matter had a beginning also, or that matter is not eternal."

From that point forward there could be little doubt that this was indeed Leitch's series of essays. It was extremely rare for authors to share authoring duties in ongoing series in the Victorian periodicals, but common for them to disguise their identity. One such author was famed astronomer Richard Proctor who wrote on many of the same topics for the *Cornhill* magazine, completely anonymously.

In the fourteenth essay Leitch tackled a subject which didn't appear in his later work; a topic which still haunts everyone today; whether machines have the ability to replace people. In this short extract we see both his familiarity with Babbage's first attempts at building a computer and the long-term prospect of machines having a "moral" aspect; something not very far removed from today's discussions about artificial intelligence.

"If man be viewed economically as a productive agent, he may be taken as made up of three elements—strength, skill, and character, each of these bearing its distinct commercial value in the labour-market. Machinery, realizing the idea of the monster in Franken-stein, comes into the field to compete with him on all these points.

The steam-engine, in its various applications, laughs to scorn its puny antagonist. Its iron skeleton needs not the luxuries which pamper the human frame. It asks only four pounds of coals daily, to match the strongest Navie, and it will work just in proportion to its consumption, so that with the wages of one Navie, it will do the work of one hundred and twenty. The triumphs of machinery, in point of skill, are, perhaps, more wonderful. Our manufactures supply innumerable examples, where wheels and pinions perform work, for which skilled labour was formerly employed, at an exorbitant price. The nearest approach to human intellect, is the celebrated difference-engine of Babbage, or rather, the analytical engine in which he is now at work. By this latter machine feats may be performed, evincing skill, equivalent to years of mathematical training at our colleges. It might be supposed that, although machinery competed with man as a physical and intellectual agent, it would have no chance with him as a moral agent —such, however, is not the case."

This familiarity with Babbage and his remarkable early computer designs also connects Leitch's opinions to another article he wrote that same month which he called *Mind and Matter*. In this treatise we discover that Leitch's opinions on the special entwined nature of light and time (see page 112) were in fact ideas he had been cultivating since at least 1849! Leitch was reviewing a book which had just come to his attention. The book was called *Stars and the Earth* and was first published in Germany in 1846 and then followed by the unauthorised English edition which had just appeared in 1849. The English edition was unattributed, but it was originally written by Felix Eberty, a German mathematician and astronomer. What Eberty did in this short book was described how a voyager moving into space faster than the speed of light could look back on the past. Although reviewers of the time generally seem to have applauded Eberty's original thinking and most modern scholars now credit Eberty with lighting the fire under Einstein, Leitch was less impressed.

In his review he called it "a small work with great pretensions" and he dismissed the "breakthrough" comments about the strange relationship between light and time with, *"The startling fancies that may be based upon this property of Light have been familiar to all who have turned their attention to optics."* Leitch was so convinced that this was not a new revelation he went on to state that the main *"claim of the work...lies wholly in the absurd attempt to explain the Divine Omniscience."* Eberty had suggested that because every light ray from Earth propagates into space forever, he deduced that this was how God was aware of everything that individuals have done in their lives and can thus keep a record of their sins. Leitch's review is scathing and in places amusing and uncharacteristically sarcastic.

"Do the reflected rays of the sun really keep a record of all that comes under the Omniscient eye of God? The author makes a ludicrous admission which ought to have awakened him to its absurdity. He says that no record could be kept of what occurred under the roofs of houses—that all this must be omitted from the supposed condensation of the world's history. We suspect if the roofs were lifted off, a far wider and stranger history would be presented than what is furnished by the deeds of men in the broad light of day."

Remarkably Eberty's book, despite originally appearing anonymously, survived long enough to have an edition published in the early 20th century with a foreword by none other than Albert Einstein who called Eberty *"an original and ingenious person."*

So despite this new insight we are still left with the conclusion that Leitch understood light's special properties before he discovered Eberty's small volume in 1849, even though this was destined to be one of the books which inspired Einstein decades later. For once the cards seem to have fallen briefly in Leitch's favour since *The Reader* magazine of 31 January 1863 gave a very positive review of Leitch's book and drew particular attention to his comments on the relationship between light and time which had appeared in *A Journey Through Space* describing it thus, *"Let the following extract from the first chapter suffice to show the **vein of originality** which pervades the book..."* This suggests that the reviewer had never seen anything like it previously, despite Eberty's earlier efforts.

Another essay that month was entitled *Herschel's Observations at the Cape* and it was soon followed up with one named *Nebulae*. These two essays must have made up the bulk of Leitch's first well-documented public science lecture which he gave in March of 1850 ^{(see} ^{page 33)}. The essays include an elaborate explanation of how the nebulae seen by both Herschel and Rosse seemed to follow the same dynamic rules as the rings of Saturn (i.e. they were thought to be rotating spirals, still awaiting confirmation as such.) Leitch also talked in some detail in this essay about the methods of observation and making recordings with automatic recorders [see page 120].

In Part 4 of his string of essays for MacPhail he further developed his argument which would appear again in *Good Words* in 1860. The belief that just because a new scientific discovery seems to contradict the Bible it is not cause to dispute it [see page 85]. In this case he was citing the apparent age of geological and astronomical formations.

"... we are guided in our belief not by possibilities but by probabilities. In the astronomical as well as the geological argument, no one can deny the required possibility, but no one with a sound judgment will accord his assent to the improbable supposition. Is it wise to clog the evidences of Christianity with such overwhelming improbabilities? When alleged facts in science are made the basis of dangerous speculation, there are two courses open to the friends of divine truth, either to dispute the facts, or if indisputable, to show that the speculation does not flow as a legitimate sequence from the facts."

Somewhat surprisingly Leitch gently berated David Brewster who had said that the nebular hypothesis of planetary formation was wrong because no one had ever witnessed it taking place. This is where (for once) anonymity perhaps helped Leitch, since he was well-acquainted with Brewster at this time [see page 43].

One peculiar thing seems to be that in this essay he also illustrated that he was already aware that the contentious Bode's law was somehow flawed, and yet he still chose to include it in the back of his book thirteen years later [see page 83].

In his essay titled *The New Force* he spoke at considerable length about the amazing discoveries of Faraday, including the great scientist's work on the strange properties of crystals and piezo electric charges. This led to Leitch contemplating what it would take to solve the biggest mysteries of the universe. This is, for all intents and purposes, Leitch talking about a "theory of everything."

"It has yet to be shown whether the higher problem now presented requires in one mind the combined elements of a Newton and a Faraday. The recent advances of science have held out strong hopes, that the imponderable elements, light, heat, and electricity, might, along with gravitation, be resolved into some higher unity."

In these essays he also included quotes from his old teacher John Pringle Nichol and makes at least one citation from the Greenock newspapers (where he grew up.)

Not all of the essays were about science; after all, the series was supposed to also be about theology. We have seen that later in life he was less than amenable towards the Catholic Church, but in these earlier essays his opinions on the "opposition", i.e. Judaism and Catholicism are much more strident. In 1849 The Catholic church was only just beginning its rehabilitation after more than a century of being portrayed as the enemy. Leitch slips into full evangelical mode and addresses everything from the science of Christ's death on the cross due to blood-loss to what can scarcely be called anything other than diatribes against Catholicism and Judaism. Seen from the lofty vantage point of the 21st century his comments on Judaism are overtly offensive, but they are not atypical of many of the distressing opinions of the time. Despite these prejudices, Leitch did at least recognise that a reconciliation between Christians and Jews was long overdue and that the acts of some Christian leaders had actually precipitated the misfortunes of the Jews and were just plain wrong, while in the same month he also delivered an extraordinary explanation of the history of the electric telegraph in which he gave full credit to a "Jesuit", for essentially conceiving of the device in 1617. This seems to reinforce his stance that regardless of your background or faith, if you are the discoverer of a "fact" you should still get the credit. This led Leitch to ponder the way that history never remembers the first person to think of something, but only the first to actually make use of the idea.

"This affords one of the questiones vexatae *which every now and then turn up in the history of science, in reference to priority of invention. The credit is ascribed by the public, and naturally so, to the man who makes an invention practically available for useful purposes. We have in this case of an illustration of the fact, that inventions soon perish when born prematurely—when the world is not prepared to appreciate them. In Egypt, there is little doubt, that many optical and other scientific discoveries were made, but soon perished, as the immediate application to useful purposes was not obvious. It is the commercial spirit of the age that at once appropriates new discoveries and inventions. Science is a power—but it is a power for both good and evil. All depends upon the spirit which wields the power."*

Little did he know he may as well have been describing his own ideas about rocket space flight.

Another of his favourite subjects was tackled in one of these essays; the conflict between secular and church-run schools. He called it simply *The School* and it is clear from this 1849 assessment that his ideas were just coming to the boil. This article appeared at almost the exact time that the Church took its more enlightened posture (see page 57). It may also explain why so many didn't believe him when he argued in his later life for equal rights for both kinds of schools (see page 58). Evidently his opinions on this most contentious topic mellowed over time.

In yet another essay that month, entitled *Philosophical Errors*, he spoke about how dogma can slow progress in both scientific and spiritual pursuits, and to prove his point he discussed the "luminiferous ether." Ever since Newton's day scientists were struggling to understand how gravity could propagate across a distance if there was no "medium" in which to transmit its effects. Thus the popular solution was the invention of an invisible "ether". Everything in this statement remains consistent with Leitch's opinions later in life, about being willing to embrace new ideas rather than insulate yourself from them. It also shows a glimmer of his later opinions about the creation of the universe (see page 133).

"In astronomy, (dogma) necessitated the invention of an ether by which all the bodies of the system might be connected, and through which gravitation might be propagated. But the assumption of this ether does not in the least render the action of one body upon another intelligible. The mystery is the same, either on the supposition of an ether or an absolute vacuum. The undulatory theory of light renders the idea of an ether plausible, but no a priori *argument from its comprehensibleness could establish it. On the subject of creation we hear, many of the ancient theological and psychological aberrations can be traced to the inexplicable nature of absolute creation. It was admitted as an axiom, that there could be no such creation—that the substratum from which all forms of mind and matter were reduced, existed from eternity. On the spiritual side of the question, this naturally landed in the doctrine of metempsychoses. A man's spirit being eternal, not only his present form of existence must be accounted for, but his past history. Hence the idea of the transmigration of souls. Viewed in this connection, we cannot be surprised that the doctrine should be so*

widely prevalent, wherever the human mind enlightened by revelation has attempted to solve the mystery of man's being. Absolute creation, as revealed in Scripture, is inexplicable, but any hypothesis devised to evade it is equally inexplicable, and as we have seen, necessarily lands in absurdity."

Perhaps more importantly we also get a glimpse in this same essay of some of the ideas which ended up in the first and most important chapter of his book, the essay *Journey Through Space* ^(see page 94). He explained in some detail in the 1862 book that the observer's vantage point is critical to understanding the mechanics of the solar system. Here it is in 1849:

"The eye of the observer is very unfavourably situated for understanding the movements of the bodies of the solar system; his eye being in the plane of their motions. The disadvantage is similar to that which we should feel, if, in studying the movements of a game of chess, we lowered our eye to the level of the board instead of looking right down upon it. The observer, if a stranger to the constitution of the solar system, would soon discover anomalies. If told that all the planets moved in one direction, he would at once point to their introgradation as contradictory facts. If, however, he could by some means leave the plane of the ecliptic and occupy some position far above it, the unity of the system would at once flash upon him and reconcile all the apparent contradictions, for then the planets would be seen moving in concentric orbits, in perfect order, and all in one direction. In regard to astronomy, man for some thousand years occupied the unfavourable point of view, and was therefore perplexed with many riddles. We, in these latter days, have been privileged, by the aid of Copernicus, to occupy a favourable point of view, from which all the riddles are solved."

And if the reader is still not convinced that all of this was by William Leitch (after all, it is unattributed) we come to the essay published as part of the series in April of 1850 called *Electrical Storms*.

This article appeared swiftly on the heels of Leitch's correspondence with Lord Kelvin about the lightning strike at a local house ^(see page 76). This event took place at Hahill Farm in Collessie just half a mile from his home in Monimail. In Lord Kelvin's diaries we only get a glimpse of just how much work Leitch put into analysing the damage caused by the storm, but in this essay we get the location, and an immense amount of detail. This description could only have been made by Leitch and still remain consistent with the facts recorded by Lord Kelvin.

In this commentary he also expressed how scientists were beginning to understand just how much energy was locked up in the atom and its destructive potential.

"The elements of nature are like a magazine of gunpowder, which only requires the igniting spark to destroy the equilibrium of the pent-up forces. Every mouthful of water we swallow contains a thunderstorm. There are condensed around its atoms as much electricity as would suffice for the most terrific storm that has ever made the heart of man to quail. Faraday has established the important law that the amount of electricity necessary to decompose a certain quantity of water is exactly equal to the amount of electricity liberated in the decomposition. The first can readily be measured, and, consequently, the second is at once ascertained. It has been found in this way, that the atomic elements of a single drop of water, viz. hydrogen and oxygen, are held together by an amount of electricity equal to a thunder peal."

Perhaps of incidental interest is that he also spoke about the storms which occasionally seem to envelop the earth. He described just such a global storm on 24th September 1841 which was *"observed simultaneously at Greenwich, Prague, the Cape of Good Hope, Van Dieman's Land, and Macao. These violent oscillations thus indicated a storm which wrapped the whole globe in its influence."* This was Leitch recording an event that was the beginning of an almost 30-year long arc of intense solar activity.

This concludes the summary of his contributions to MacPhails. We now move on to the new material obtained through other sources.

One of the most difficult things to conclusively prove was that Leitch taught at Glasgow University ^(see page 28). Even the University's own records are incomplete, but more evidence has now come to light. In December 1859, in a report to the trustees of Queen's University, reprinted in the journals of the time, there is a reference to Leitch teaching in Glasgow "during the illness of a professor" and about his "eminence in Astronomy and the Natural Sciences." The same month the Halifax Record reprinted the story which stated:

"When at Glasgow University, nearly thirty years ago (sic), he was one of our most distinguished students; so much so that when the Professor of Astronomy was unable to attend to the duties of the class, Mr Leitch was appointed to fill his place for the whole session... He has ever been one of the ablest contributors to MacPhail's Journal and other periodicals, and, whether on the evidences or on questions of statistics or church polity, his writings show a clearness, a readiness, and a grasp of argument that very few in Scotland can equal...I only wonder that they were able to offer him sufficient inducements to leave Scotland."

This seems to suggest Leitch not only filled in for Meikleham, but did so for some considerable period of time while William Thomson was still a student in Meikleham's class.

Beyond the fact that this ties Leitch's tenure as a teacher to the illness of Meikleham it also shows that Leitch was already well-known for his magazine articles before MacLeod and Strahan launched *Good Words* the following month.

Five years later *The Presbyterian* reported that, *"While a student he also lectured in the University on Astronomy, and for several years acted in the Observatory connected with the College, as assistant to the eminent astronomer, the late Professor Nîchol."*

In August 1864 the same journal reported that the Reverend McNair *"knew him at the time he was gaining his honours at Glasgow University, where he was regarded as one of the first students there. So high was the opinion he occupied as a scientific student, that he was appointed lecturer one session on his favourite study of Astronomy."*

The Presbyterian proved to be a rich mine of information about Leitch and in November of 1860 it also reported his arrival in Canada noting that his new book would be published soon, *"...chapters of which have already appeared in Good Words and have been perused with lively interest by many readers in Canada."* This comment seems to confirm the wide-ranging distribution of Strahan's monthly magazine, even in its first year of publication.

On December 7th 1860 Leitch addressed the first meeting of the Botanical Society of Canada by saying, *"Universities do not discharge all their functions by merely teaching the acknowledged truths of literature and science; it is a part of their duty to organize and insti-*

gate original inquiry in the different departments of knowledge." He would later reiterate this and similar opinions to the student body at Queens. Leitch also earnestly believed that the marvels and assertions of science had no meaning to the "everyman" without an explanation of the methods used by scientists to arrive at their conclusions. He felt it was not enough to just spout data; a lesson which some experts could learn from today.

Around the same time as his address to the Botanical society he made his first official address to the university faculty and students. In this speech he made references to James Watt, which is hardly surprising considering the amount of time he had spent at Glasgow University, but it does confirm his first-hand knowledge of the pioneering engineer who changed the world. Leitch very pointedly also again used the expression *"action and reaction"* in his speech, although more metaphorically than specifically in reference to Newton's revelations. In one of the MacPhail essays he used it to discuss Newton.

In January of 1861 Leitch wrote a letter to the school trustees explaining how the school could save money by using the various available heating options more efficiently. The letter doesn't exhibit any unprecedented genius, but it does demonstrate that Leitch was using his experiences in Scotland, where he had invented a new kind of heater, to help improve things in his adopted country.

The following month Leitch wrote another letter to an undisclosed recipient in which he demonstrated his penchant for diplomacy. The so-called "University Question" was in full swing and he expressed his opinion that he didn't want to stir up trouble, but that he still intended to take his seat in the University of Toronto Senate. He would later have to decline from taking part in the U of T examinations because of his return to Scotland that summer.

On the 8th March 1861 Leitch gave a long lecture to the Botanical society entitled *On The Sexual Development And Economy Of Bees, And on the Saccharine Matter of Plants, Viewed In Relation Thereto.* Although it is still unclear how many academic papers and lectures that Leitch gave in his life, besides the aforementioned series in MacPhail's we have copies of his papers on plants and insects; at least one of his lectures on astronomy (included in the following pages) and of course his book. We also know that somewhere he wrote or lectured on ballistics and that he taught mathematics, ethics, divinity and philosophy. He was, by any reasonable estimation, a polymath.

In April of 1861 Leitch gave a lengthy speech to the *Missionary Society of Queen's College*. Many of his talks to the student body were about the missionary purpose of the church, not only to spread the gospel in Canada but around the world. When he was appointed by the synod to go to Canada it was made quite clear that teaching the people about the church was just as important as any of his other roles.

While in Edinburgh in August of 1861 Leitch wrote a letter back to Queens about buying observatory equipment and books for the school. One of the first things he had done on arriving in Kingston was to establish a professional observatory. The local facility was owned by the municipality of Kingston, but he had brokered a deal whereby the town would hand the observatory over to the college and finance the purchase of better equipment. In exchange for this beneficence the town had delivered their terms, which included allowing the city elders to use the facility whenever they wished; allowing distinguished citizens access for a small fee; and making sure that the observatory had a person employed at all times to set the local town clock. They also demanded that a regular series of lectures be provided to the townspeople. Leitch had his own ideas of how to accomplish all of this. He even sketched out his design for the instrument layout in the new observatory. Parts of his letter are illegible but a best-guess transcription is provided here alongside the original.

My Dear Sir,

I hope you are progressing with the observatory and that the room will be ready for the winter lectures. I visited the observatory of Edinburgh two days ago and had the opportunity of examining all the arrangements. I shall lend you a report with a detail of the plan adopted for fixing time to the time. The plan is very perfect and could be easily adopted in Kingston. Any clock in the town of Kingston - say that at the post office might with very little expense be made to beat seconds in perfect coincidence to the clock in the observatory. This was often before tried by making the secondary clock go by electricity, but it always failed. In the new plan the clock goes by weight as a common clock and the electricity, which is very weak, is employed merely to regulate. All

bought to a greater perfection. I am
to go over to Endulyts soon to see
it tried

I have ordered the prizes
for last year and also for
the ensuing year, as required by
Mr Drummond. I felt at a loss
what books to order, but I have
my rank orders the same books
as last year. In your class instead of the class books of last
year, I have taken those you have
given in the Prospectus, as I thought
it would be your wish to give the
books mentioned there.

I have ordered books nearly
to the extent you wished us also
They will be sent in

I remain & truly

Yours ever
William Smith

My Dear Sir

I hope you are not uneasy
about the observatory and that
the room will be ready for
the winter lectures. I visited
the observatory of Polubrsk
two days ago and had the
opportunity of examining all
the arrangements. I shall send
you a report with a detail
of the plan adopted for fixing
down to the time. The plan
is very perfect and could be
easily adopted in Kingston

---

any Clock in the Town of Kingston
– I say that at the best offices
might with my little scheme be
made to beat seconds in perfect
correspondence with the Clock in the
Observatory. This was often before
tried by making the secondary
Clock go by electricity but it always
failed. In the new plan, the
Clock goes by weight as a common
Clock and the electricity which
is my weak is employed merely
to regulate. All that is necessary
is to get a pendulum for the secondary
Clock which may be of itself an
imperfect Clock. I am ordering
some Maps and apparatus for the
Observatory. The Single Thought

should not less than 30 feet hence
be required from south to north line
due in transit room. to admit of
horizontal Collimations

North Collimator   transit   South Collimator

not less than 30

There should be as few windows as possible
so that the sun may not shine when
the instruments or pillars. I should
like the North left a photo electric
Microscope, which would be very
useful in observing polarisation and
other astronomical phenomena but it
is a little beyond the mark
I am not sure also whether it is

that is necessary is to get a patent pendulum for the secondary clock which may be of itself as any imperfect clock. I am ordering star maps and apparatus for the observatory. The length ??? this not less than 30 feet should be required from south to north window in transit room to admit of horizontal collimators. There should be as few windows as possible so that the sun may not shine upon the instruments or pillars.

I should like much to get a photo electric microscope which would be very effective in showing polarisation and astronomical phenomena but it is a little beyond reach. I am not sure also whether it is brought to requisite perfection. I am to go over to Edinburgh Sun(day) to see it tried. I have ordered the prizes for last year and also for the ensuing year. As requested by Mr. Drummond. I felt at a loss what books to order, but I have very much ordered the same books as last year. In your class instead of the class books of last year I have taken there your leave given in the prospectus, as I thought it should be your ??? to give the books mentioned there. I have ordered books nearly to the amount you mentioned or £40. They will be sent out shortly. I remain

Yours very truly

William Leitch

In keeping with his missionary role, on June 4th 1861 Leitch gave a speech to *The Ladies Association For Female Education In India*. Still further evidence of his enlightened attitude toward equal opportunity is found in another letter in which he stood up for a deaf applicant who had applied for work at the College. His letter on that occasion clearly stated that he saw no reason to stymie the man just because he was deaf. It can be seen from both his actions and the endorsements of his colleagues that Leitch actually tried to live by the Christian doctrine.

On 18th October 1861 during his long sojourn to Scotland Leitch sent another letter to Queens. In this he announced his plans to donate a precision clock to the observatory along with his 6½" Shortt reflecting telescope (made in 1742). This expensive donation was used as a carrot to persuade the local government to also start chipping in more money. Leitch also ordered an oxy-hydrogen microscope for the school. This peculiar piece of equipment is described in detail in a book called *Micrographia* published in 1837 (see diagram below). By combining the two gases and igniting a small piece of lime an extremely bright light could be generated for observations with the microscope. Of course combining gases in blow-pipes had been around since antiquity but the discovery of oxygen and hydrogen in the 18th century had led to a series of new uses for the technology.

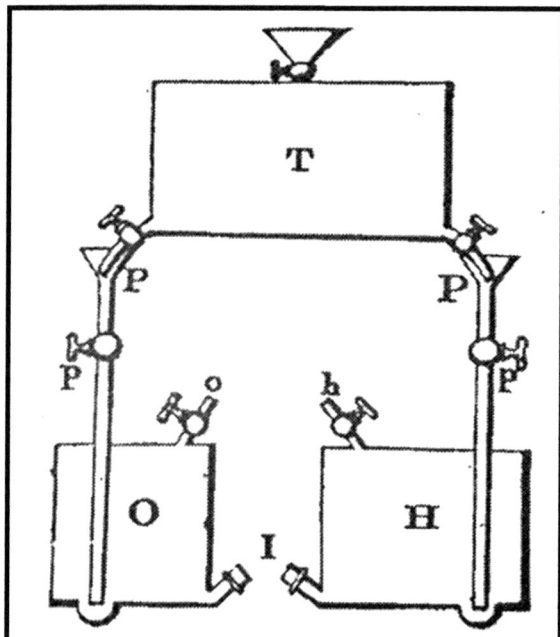

In the diagram at left, T is a water tank providing gravity-fed pressure to two valves (P) which then force oxygen and hydrogen (O and H) out of two more valves (o and h) "exit pipes fitted to communicate with the jet where the ignition is to take place". The two orifices at letter I, are used for filling the gas tanks.

If you add a combustion chamber at o and h you've got a very inefficient rocket. Evidently the oxy-hydrogen microscope became quite common at this time and supplanted all daylight microscopes. What is also of passing interest to space historians is that Erasmus Darwin, who was acquainted with Joseph Priestly, the discoverer of oxygen, sketched something very similar in his so-called "Commonplace" book in 1779. The Darwin museum in England suggested that this was an oxygen hydrogen rocket, which, if correct would of course change history, but to this author it looks suspiciously like an early oxy-hydrogen lamp. From this letter and the essay in MacPhails we can see that Leitch was well-acquainted with the combination of gases that would one day propel men to the moon.

In November of 1861 Leitch decided that he was going back to Canada to stay permanently. This was reported with enthusiasm in

The Presbyterian that month. The following month's issue of the same journal reported that he had formally resigned from his post in Monimail/Cupar, and one of the reasons he gave was, *"The creation of a National Observatory in connection with the University will enable me to turn to account my early training in Practical Astronomy by directing the course of observations, and making the cultivation of astronomy subservient to the moral and religious elevation of the people."*

On January 10th 1862 after the meeting of the Botanical Society the minutes stated, *"Principal Leitch spoke of the importance of such researches in several points of view. The fact of plants being found to inhabit definite zones or lines along the shore to which their distribution was restricted, as found by Mr. Kemp, served to show that there was here an apparent barrier to that tendency to specific change which is argued for in the speculations of Lamarck, the author of the Vestiges, and Darwin."* Just two years after Charles Darwin's paradigm shifting book, Leitch was applying his discoveries to understand different ecological habitats.

After returning from his first trip back to Scotland Leitch now had to keep up his end of the bargain with the people of Kingston. Adverts were placed in the local newspapers announcing the first in the contracted series of astronomy lectures. On March 28th 1862 Leitch apparently dazzled the local populace at Kingston City hall with his gifted oratory. His second lecture a few weeks later precipitated the following review in *The Presbyterian,*

POPULAR LECTURES
ON
ASTRONOMY
BY THE
REV. PRINCIPAL LEITCH, D. D.,
OF QUEEN'S COLLEGE,

BEING the first of a series of lectures under the provisions of the Deed of the Kingston Observatory.

The lectures will be delivered in the CITY HALL, on Friday Evening, the 28th instant, and Friday Evening, the 4th April.

His Worship the Mayor in the chair.— Admission free. Each lecture to commence at half-past 7 o'clock.

The subsequent lectures will be delivered in the Observatory, accompanied by actual telescopic illustrations, and of which due notice will be given.

Kingston, March 26th, 1862   72

"Lecture On Astronomy -The Rev. Principal Leitch delivered the second and last public lecture for the

season in connection with the Observatory Trust Deed on Friday night. It was an interesting indication to see the City Hall crowded to excess, numbers being compelled to stand, and many to go away without being able to obtain admission. Such eager attendance at a scientific lecture is very unusual though it must be confessed the lectures are rendered very attractive by the exhibition of the magic lantern apparatus and other illustrations and experiments. A very full synopsis of this lecture will appear in a subsequent issue."

Around the same time Leitch delivered an inspiring speech to the newly graduating Medical students at the college. He urged the students to investigate the arts so that they might have a better-rounded education; to be compassionate to their patients; and to do credit to their teachers.

In the summer of 1862 Leitch was made moderator of the Synod in Kingston. During the same month he received a good review in *The Presbyterian* for his writings in *Good Words*. At this point most of what he had written for the magazine was now in print. All that remained would be saved for the book which was finally in Strahan's hands and being prepared for publication. There was one notable exception, a lecture he had written expanding on his theories about the nature of the sun. In September, just before the publication of his book, the sun lecture appeared in the pages of *The Presbyterian*. In this extraordinary talk Leitch demonstrated his familiarity with spectroscopic analysis, gravity, Newton, calculating the mass of distant objects, parallax, renewable energy, kinetic energy and the mortality of the solar system. Here it is in its entirety.

> It is remarkable that the most important body in the solar system should be the one whose physical constitution and structure attracted, till lately, least notice. It seemed hopeless to fathom the mystery of this fountain of light and heat. The milder rays of the other bodies of the system allowed us to gaze comfortably on their surface, and to trace resemblances to our own globe; but the sun repelled us by his fierce rays, and astronomers contented themselves with a rapid glance, as if looking into a scorching furnace. The sun was regarded as wholly dissimilar to the other bodies of the system—so dissimilar indeed that it was thought no knowledge of terrestrial conditions would ever

enable us to comprehend conditions apparently so different. What difference could be greater than between a fierce furnace, like the sun, and cold, dark, solid bodies, like the planets? The sun appeared to be a mystery so profound that astronomers felt it was irreverence to pry into it too curiously. Recent science has, however, thrown off all delicacy on this subject, and the sun is now treated as familiarly by the chemist as any substance submitted to his analysis. It has been found that the sun is not wholly dissociated from the planets in constitution and structure, that there are links of connection which show that they belong to the same family of bodies, and it is one of the chief charms of astronomy to trace these links.

The first point for consideration in discussing the subject is the measurement of the distance, size and weight of the sun. When the more startling facts of astronomy are stated to an ignorant or illiterate man, they are received of course with incredulity and, it may be, with ridicule. They so far transcend the circle of his own narrow conceptions that he smiles at the credulity of the learned. Now this incredulity is not confined to the ignorant and illiterate. Well educated people have often a secret unbelief as to the facts of astronomy, though they may be ashamed to put their opinions in opposition to that of the whole scientific world. Yet, when told that the earth's surface spins round with the velocity of a cannon ball, that the little prominences that can be seen with the naked eye on the edge of the moon are vast mountains, that the earth is no more to the sun in magnitude than a single stone of St. Paul's to the whole fabric, they are inclined to shake their heads, although positively assured of the facts by the most eminent astronomers.

They read books which give facts and figures, but still they do not bring conviction. And why does this secret unbelief cling to the mind? Simply because we do not understand the rationale or principle by which these astounding facts have been arrived at. If we once comprehend the method, the facts will readily bring conviction. Now it is this comprehension of the methods and principles of a science that constitutes real scientific knowledge. It is not the storing-up in the memory of the facts and figures of astronomy. A clever boy at school will, in the course of a few months' study, become a more profound scholar than Newton or Herschel, if astronomy consists merely in the recollection of its facts. In company Newton sometimes appeared more ignorant than others about his own discoveries, simply because he had not a memory for numbers. And some, who could not in the least comprehend the science, yet appeared in conversation to be superior, because they could at once give the exact distance of the moon or the exact compression of the earth.

In order to derive true enjoyment from the study of astronomy, and really to believe in its facts, it is necessary that you clearly comprehend the methods by which these facts have been arrived at. But you will ask, Is it possible for the popular mind without a special technical training to attain this? I think it is. It is not at all necessary to comprehend the principles of the celestial mechanism that you should be able to handle astronomical instruments or manage mathematical formulae. It is just like understanding the principle of a steam engine. It is not necessary, that you should be a practical engineer, and able to calculate the pressure of steam or the strength of materials, to comprehend the principle on which the engine works. So in the celestial mechanism you may have a thorough comprehension of the general principles involved, although you cannot enter into the technical details of calculation. In determining the distance of the sun the astronomer only employs a principle which you daily take advantage of in estimating distance. On looking out from the windows of a railway carriage you observe that near objects flit along the horizon, while distant objects creep very slowly, and you calculate that the slow objects are more distant than the fast ones. The distance is in direct proportion to the slowness of the motion. If the near house is one mile distant, then you conclude that the more remote one in the same line is two miles if its motion is twice slower; three miles if thrice slower, and so on. You have only to measure the distance at the first house, and the distance of the farthest off is at once known by ascertaining its comparative rate of motion. Instead of the most distant house you may take a cloud, or the moon, or any heavenly object. The principle is precisely the same, only you must move farther to see any appreciable change of place. This change of place according to the different position you occupy is called parallax, and on this depends on your knowledge of the distance of the heavenly bodies.

When you have in this way found the distance of the sun it is easy to measure its size you can do so by the rule a simple proportion. Suppose that, when you hold but a sixpence at arm's length, it exactly covers the face of the sun, you say that the sun and the sixpence have the same apparent size, but the sun appears so much less than it in reality is, just in proportion to its greater distance; and, if you wish to know how much larger it is in reality than the sixpence, you must ascertain how much more distant it is, or how many arm's lengths there are between the sixpence and the sun; and that number will be the number of sixpences required to stretch across the sun, and, knowing the diameter of the sixpence, you know the diameter of the sun. Then, as to the weighing of the sun, this appears still more wonderful; and, when the astronomer speaks of weighing a planet, people imagine that it is only in a metaphorical sense that he does so. But he weighs the planets just as really as the grocer weighs his goods over a counter. When you put a letter into a spring balance you think it is only the letter you are weighing, but you are at the same time weighing the earth. You are not apt to think so because the world is always the same, while you change the letters. But suppose you change the world instead of the letter.

Suppose that a letter which weighs an ounce is carried in the spring balance to another planet, and held at the same distance from its centre, would the letter weigh the same? By no means; if the planet is only half the weight of the earth, the letter will be only half an ounce; if it is double the weight of the earth, it will be two ounces. Let us suppose the one ounce letter to be carried to the sun, how much would it weigh there? Eleven tons; and, just as eleven tons is greater than one ounce, so is the sun greater than the earth. It must be carefully observed in weighing the planets that the balance must be held at the same distance from their centres, not their surfaces. But you will say, How can you get the balance conveyed to the planets and the sun? The answer is that there are nature's spring balances in the heavens.

These are the orbits or circles in which the planets move. They may be compared to bent steel springs and, just as the earth by its weight or gravity pulls and bends the wire of a spring balance so does it bend the path of the moon into a curve. Were it not for this bonding power the moon would move in a straight line: but the earth bends the straight path of the moon as the cooper bends the hoop of a barrel, and exactly in proportion to the bending in a given line is the pulling power or weight of the earth. The weight of the sun is found in the same way. You have only to measure how much it bends the paths of the planets in a given time. Knowing the weight of the earth, we can readily tell how much heavier the sun is from its superior power in bending the orbits of the planets. Let us next attend to the position of the sun in the solar system. It is the centre or the whole. In this way its light and heat are equally distributed throughout the whole year of each planet. Each planet goes round the central fire in a circle during the course of the year. We might conceive a dark body corresponding to the sun in the centre, controlling all the planets by his gravitation; while another body, such as Jupiter, had assigned to it the function of dispensing light and heat to the solar system. But, were this the case, the various planets would experience extremes of heat and cold which would be destructive to all life. Placed as the furnace is in the centre, there is but little variation in temperature in the course of the year.

The next point of interest is the structure of the sun. The spots on his surface, which Milton so poetically represents as demons flitting across his disc, reveal the true structure. These spots are not really on the surface, but holes down which you see through the dark body of the sun. When you look down these funnels you see the edges of the concentric shells of which the outer part of the mass of the sun is composed. There is probably a solid core, and round the core there is layer upon layer, like the concentric layers of a bulbous root. These layers strata are separated from one another by intervals which are probably filled with a transparent atmosphere, just as one stratum of clouds is suspended above another by the buoyancy of the earth's atmosphere. Three distinct concentric shells have been discovered by carefully looking down these abysses, which are so large that our earth could easily be projected through them. These strata are not solid, for you see the whole mass in commotion like a boiling cauldron, and its continuity is broken by these openings or holes, which are like breaks in the continuity of a cloud-covered sky. The dark body of the sun appears through them as you see the blue sky through breaks in the cloudy stratum above us. The outer visible stratum is called the photosphere, as it is from it that the light comes. Total eclipses reveal a new stratum which at other times is quite invisible on account of the brighter radiance of the photosphere. It is of a rose-coloured tint and envelops the photosphere. We are in fact looking through it, when we are looking at the bright disc of the sun. The veil is, however, so transparent that we do not suspect that we are looking through it. When the moon in a total eclipse entirely covers the sun, this rose-coloured stratum shines out with very lofty prominences, like the crests of waves in a storm. This rose-colored stratum projects only a very little beyond the limit of the sun, but in a total eclipse there is a corona like the glory round a Saint's head, which extends far beyond the limb. There is still great doubt as to the nature of this corona, whether it belongs to the sun or moon, or is merely an affection of light in passing the edge of the moon there is however no doubt that the red flames belong to the sun. The sun is encircled by rings of zones, corresponding to the rings of Saturn. The rings of Saturn cannot be solid, as was once supposed, at least they cannot form a rigid mass like a rock. They are probably composed of innumerable small masses of matter, each moving independently like a separate planet, but then so closely packed together that the mass appears solid.

One of these rings in fact, is composed of such fine particles of matter that you can see through it—this is the dark ring lately discovered. The others probably differ from it only in being more massive, or composed of coarser material, so that the stratum is too thick to be transparent. The sun has similar rings. The Zodiacal light is probably one of these. The zone of asteroids between Mars and Jupiter is another, for, although we have discovered only 70 distinct bodies there are probably millions more of a smaller size. Leverrier has also indicated 2 other zones, one within the orbit of Mercury, and the other near the orbit of the Earth. He has even approximated the weight of each of these zones or rings.

The next point is the work of the sun. It is not only to the heat and light of the sun we are indebted. Almost all the mechanical power on the face of the earth is traced to the sun. The sum of force in the universe is always the same, just as the sum of matter is always the same. The force may change its form, but its amount is always the same. This principle is known by the name of correlation of physical force. When the river leaps over the Niagara Falls and reaches the level beneath, its mechanical force is lost as to form, but it is transmuted into heat. The water at the bottom of the fall is increased in temperature, and were this heat collected, it would be converted into mechanical power, exactly adequate to raise the water to its former level. The heat of explosion is converted into mechanical power when the ball is impelled from a gun. The mechanical power is reconverted into heat when the ball is suddenly arrested in its flight. The ball will be found to be hot exactly in proportion to its velocity when arrested. Now this is the case with the sun's heat. All the mechanical power employed by man can be traced to the sun. The water wheel is turned by the sun. Its heat raises the water from the ocean and deposits it in the form of rain on the mountain's side. The river collects the rain, fills the buckets of the water wheel and by this process the sun indirectly works the machinery of the mill.

The steam engine is not an exception. Its power is derived from the heat of the furnace, but the furnace depends for its power on fuel. But how should fuel possess this power? It has derived it from the sun. The fuel as growing wood stored-up the power dispensed by the sun. The tree is the concentrated power of many summers' heat, and, though it may lie for thousands of years as coal in the bowels of the earth, it retains the power till it is evolved by burning. But you will say that animal power is surely different? Such is not the case. Every exercise of animal power costs some waste of tissue. That tissue is ultimately derived from vegetable matter, and the vegetable matter owes its power to the rays of the sun. Volition cannot create mechanical power; it can only direct and apply it. The only power not derived from the sun is that of the rise and fall of the tide, as far as this is due to the moon. The trade winds may also be regarded as an exception. This power is derived from the rotation of the earth, though the heat of the sun is necessary to develop the power.

The next point of interest is the combustion of the sun. It was long thought that the sun's combustion was totally different from that of all other bodies, and that by some mysterious process light and heat could be constantly given out without any loss. The principle of the correlation of physical force tends to the conclusion that there is a real loss of power; that the radiation of heat is like the pouring of water out of a cistern, and that, unless there are some means of supply, it must be exhausted. What is more, recent science has actually discovered well known substances in the incandescent atmosphere of the sun, bringing the flame into close analogy to terrestrial combustion. The following metals have already been detected in the state of vapor in the incandescent atmosphere of the sun:—Sodium, potassium, magnesium, iron, chromium and nickel.

This has been accomplished by means of what is called spectrum analyses. The general principle is readily understood. It is the use of color as a test. You can often judge, simply by the color, as to what the nature of any substance is. When certain substances are put into the flame of a lamp, you can guess at the nature of the substances by the color of the flame. (Flames were exhibited of different colors, produced by the mixture of soda, potash, lime, strontia, with spirits of wine.) And by merely marking the shade of color you might form a good idea as to the substances which tinged the flame. Still this test would often fail, as the same color may result from the mixture of various substances. There may be various substances in the flame giving one compound color, and from this one color it would be impossible to discover the various substances. When, however, you view the flame through a prism with proper precautions, admitting the light only through a narrow slit, you find that the spectrum or colored image of the flame of each substance has a distinct pattern—has so many colored bands running across it with dark intervals between. Each substance is known by the color, number and position of the bands. If there are incandescent substances in the flame, the patterns of both are given, so that they may be at once distinguished. If the flame is supposed to become a solid, white, incandescent body, such as platinum, you get a spectrum with all the seven primitive colors, and they are quite continuous. There are no dark gaps, because the light is pure white, and comes from a solid body. There are dark gaps in the spectrum of a flame charged with incandescent particles in it, because the flame has not all the colors of white light. The sodium spectrum has only one yellow band, and all the other colors are wanting. Lithium has only a yellow and an orange band, with all the other colors wanting; and there is a dark gap between these two colored bands, because the intermediate shades of yellow and orange are wanting. The delicacy of this test transcends immeasurably all other tests. The thirty-millionth of a grain of sodium can be detected in a flame. If a bucketful of salt were thrown into Lake Ontario, and equally diffused, it could be detected in a bucketful of water drawn at any part of the lake. But how does all this bear on the chemistry of the sun?

How does this principle enable us to detect the substances in the solar atmosphere? It has been stated that a solid, white, incandescent body gives all the several colors with their innumerable shades. The sun gives this; and, if this were all, we would be entitled to conclude that the illuminating portion of the sun was also solid or fluid, for a fluid comports itself like a solid. But along with the perfect continuous spectrum there is a peculiar structure. The spectrum is striated with innumerable fine black lines, not uniformly distributed, but peculiarly grouped. Every color is thus striated, just as a rainbow would be striated if you held up between it and your eye the warp of a web, the threads running along the ribs of the bow. The interest of Kirchhoff and Bunsen's researches lies in the explanation given of these dark lines. They have shown that they are the reversed spectra of the incandescent substances in the vaporous atmosphere of the sun and that they are reversed or appear dark because they are seen on the brighter background of the white solid or fluid body of the sun. According to this theory, if the solid or fluid body of the sun were obliterated, while the vaporous incandescent atmosphere remained, all the black lines would become colored with their appropriate tints, and we could recognise the patterns with which we are so familiar when analyzing the substances diffused in the flame of a lamp. This theory is verified by actual experiment. When the brighter light of ignited lime or charcoal points is placed behind the flame of a lamp, the colored patterns give way to dark lines, which occupy the same place and preserve the same grouping. The colored bands in the spectrum of the flame extinguish the corresponding colors in the spectrum of the solid source of light, and replace them by corresponding dark lines. The color of the bars of a window is not visible when you look out upon the bright sky; they appear simply as black lines. And so do the colored lines of the spectra of the various substances appear dark when seen against the brighter spectrum of the solid source of light. By carefully examining the grouping of the dark lines in the sun's spectrum, and comparing them with the known colored patterns of various substances, the metals already enumerated have been detected. You might think it impossible to single out from innumerable dark lines the pattern of a certain metal, but the chemist can do this as readily as the sailor can single out the rig of his own ship from a forest of masts in the harbor. This spectrum analysis is one of the most brilliant achievements of our day, and will undoubtedly form an era in the history of chemistry. It has enabled chemistry to extend its dominion to the sun and stars.

An interesting question in connection with the combustion of the sun is, How is it supplied with fuel? for it cannot dispense light and heat with undiminished intensity unless replenished with fuel. The old theory that the comets are the sun's fuel is revived in another form. The comet of Encke is gradually approaching the sun in a spiral course, and will ultimately fall into it. And, although no tendency to this result has, as yet, been detected in reference to the planets, there is little doubt that the same fate is reserved for them. This may be caused by a resisting medium, or it may be due to the repelling force exercised by the sun, which all comets show in a striking form, and which the analysis of M. Faye has proved to be explanatory of the shortening periods of Encke's comet. It is believed that the zones of meteorites, approaching the sun in a spiral course, like that of a comet, gradually supply the sun with the necessary material to keep up its heat; and this can be done, though these meteorites be not combustible. Their arrested motion would supply an adequate amount of heat. These zones of meteorites are closing in like the rings of Saturn upon the central body, for M. Struve's observations incontestably show that these rings are stretching out to the body of the planet. This spiral tendency is also illustrated by the spiral form of so many nebulae. And no one can look at these spirals without the conviction that there is progress towards a centre. But the sun's fuel is limited, and the combustion must at last cease. The researches of the German chemists lead to the conclusion that the photosphere is fluid, not gaseous. It cannot be conceived a continuous solid. It is also probable that the region of the incandescent metals in the state of vapor is the rose-colored stratum seen in total eclipses. It will be a matter of intense interest, on the occasion of the next total eclipse of the sun, to ascertain whether the characteristic colored bands of the metals are to be found in the rose-colored prominences and in the corona.

We have seen that science has distinctly traced the doom written on the solar system. It is destined to pass away. The machine is running down. The central fire will at last be exhausted. The planets and satellites in their spiral courses will come to a standstill. But are we to arrive at the conclusion that God's glory shall no longer be manifested in the heavens? or that this system is to rush into annihilation? No, there is no ground in science for the belief that a single particle of matter will ever be annihilated; but there is every ground for the belief that the passing-away of the solar system is only one phase of some grander revolution, and that from the ashes of the present system more glorious worlds and systems may arise. All this is in perfect, almost literal, accordance with the Scriptures, which represent the heavens as passing away as a scroll. "They shall wax old as a garment. As a vesture shalt thou change them, and they shall be changed." It represents the phenomenal world as ever changing—in a state of unceasing fluctuation—while the great absolute I AM remains ever the same. It is with a feeling of regret that we detect anything like imperfection or decay in the heavens. We would fondly cling to the belief that the celestial mechanism is imperishable, while all things change and decay on earth. But why should the heavens be an exception to the rule, that every structure and organism has only certain periods of existence? We do not think the flower that blossoms but for a day less beautiful, or manifesting God's wisdom less wondrously because it has but a brief period of existence. The wisdom of God is displayed in adapting its structure to the period of its existence, whether long or short.

And so in the heavens God's wisdom is displayed in so balancing and adjusting the solar system that it is admirably adapted to serve the temporary purpose for which it is intended. The constituent elements of the flower pass away for a time from view, but only to reappear in some other form, and fulfil perhaps some higher functions; and so it will undoubtedly be with the elements of the solar system. And is there not a great and important lesson taught by this fleeting character of even the grandest systems of the universe? It tells us that we seek in vain for something immutable and eternal in the shadows of material things. Amongst the ceaseless fluctuations of material phenomena it forces us to seek Him who is the same yesterday, to-day, and forever. To confer upon matter the attribute of immutability, and to stamp upon systems the attribute of eternity, would be to make the universe God. It would be to deify matter and material things; whereas the ever-changing character of all created things—of systems of worlds, as well as vegetable and animal organisms—is designed to point to the personal, living, unchangeable God, who is in all, through all, and above all. God spoke the worlds into being, and worlds and systems are but the written thoughts of God. But we have no reason to believe that God has spoken His last word, or that worlds and systems are not still to be evolved from chaos. The solar system may pass away, as a spoken sound fades upon the ear, but it is after all only one articulate utterance of the Almighty. Are there not yet tones to be uttered, chords to be struck, far surpassing any utterances that have yet been heard? The spirit is overwhelmed at the vast period of the solar system, the millions of years that may yet elapse before it reaches its final destiny, but in a higher state of being, and occupying a loftier eminence, this vast period will be only the turning of a single page in the history of the universe. Milton sublimely speaks of the skies as of the book of God wherein to read His glory; but, after all, it is only the hornbook of the beginner. There are other books to be opened, deeper mysteries to be fathomed ; and the heavens above us are only the preface of that greater roll which is to be unfolded to us when suitably prepared by our training on earth. Let us then reverentially read this book, believing that it is purposely designed to fit us for a higher state of being, where we shall see no longer in part, but when with open face we shall behold the full glory of God.

On 24th October 1862 Leitch was back in Edinburgh and writing a letter stating his intention to leave for Boston on the Cunard liner *Europa* on November 1st. As soon as he returned to Canada he sent his staff on a quest to borrow an expensive transit instrument for the observatory from the Royal Astronomical Society in London. Getting this piece of equipment required intervention from some highly placed people in the government. It was listed in the RAS inventory alongside the gold-standard of scientific instruments owned by the RAS in Greenwich including Herschel's telescope and Harrison's clock. It had been built by Cary of the Strand in London, a family business also known to have made the telescope used by Lewis and Clark. It was donated to the RAS in 1827 by Lieutenant George Beaufoy when his father, astronomer Colonel Mark Beaufoy died. The Cary/Beaufoy transit was four-feet long and was still in Kingston as late as 1885. It has since disappeared.

Within five months of the release of Leitch's book a full page advert for it was placed in concurrent issues of *Canadian Naturalist and Geologist* magazine by his Canadian publisher. Within another three months it was highly placed in local book shop advertising. Perhaps this review in *The Presbyterian* in February 1863 carried some weight:

GOD'S GLORY IN THE HEAVENS: By William Leitch, D.D. Principal and Primarius Professor of Theology, University of Queen's College. Dawson Brothers, Gt. St. James St, Montreal.

The work before us, several chapters of which have already been published in "Good Words," is one of considerable interest. It treats chiefly of the higher questions of astronomy, and gives the reader a full idea of how these questions are discussed by the foremost thinkers of the day. Nor is its learned writer merely a retailer of other men's ideas; he thinks for himself and maintains and illustrates his opinions with considerable ability. He also writes in a very transparent style - his thoughts shining through it as pebbles through a running brook:—while mitering keenly into the poetry of his sublime subject, he at once enlists the enthusiasm of the reader on its behalf. These are the qualities in a writer which can render science popular; and, though some may be apt to suppose that Principal Leitch is superficial, because he makes everything so plain and simple, this is far from being the case. Many of his chapters, both in the arrangement and the matter, must have cost him much patient labour and thought. The following description of a total eclipse of the sun will illustrate the elevated style he can command, when his subject calls for it, and shows as well the peculiar mental phenomena which such a rarely witnessed event calls forth.

"It is however, when men are massed together that the finest opportunity is afforded for watching the effect of an eclipse. Such an opportunity was enjoyed by the French astronomers, when observing the total eclipse of 1842 at Perpignan. The observers were stationed on the ramparts with their instruments, the soldiers were drawn up on a square on one hand, and, on the other, the inhabitants were grouped on the glacis, so that the station commanded the full view of twenty thousand upturned faces. The astronomers did not fail to watch the phases of feeling in the crowd, as well as those of the eclipse. The moment the people with smoked glasses to their eyes, marked the first indentation on the sun's disc, they raised a deafening shout of applause much in the way in which they would salute a military hero, or a popular actor. The moon gradually crept over the sun; and, for a considerable time, there was nothing observable but

the ordinary loquacity of a French crowd. As the eclipse drew towards totality, the murmur of twenty thousand voices rapidly increased - each telling his neighbour of the strange feelings coming over him. Suddenly, the last filament of the sun's disc was covered, and, at that moment, a deep, prolonged moan, as from one man, arose from that vast crowd. It was like the stifled groan of the multitude witnessing a public execution, at the moment that the axe or the drop falls. The moan however did not mark the climax of high-strained feeling. The dead silence that ensued was the culminating point. Not a whisper was heard, not an attitude was changed as, with the rigidity of a statue, each man stood and gazed upwards. So unearthly was the silence, that the beat of the chronometers was heard with painful distinctness. The heart of the universe seemed to cease its throbbings. Nature had fallen into a state of syncope. For two and a half minutes this dreadful pause continued. At the end of this period a thread of light burst forth; the tension was at once relieved, and one loud burst of joy rent the heavens. The people could not restrain their transports of happiness, now that the dread, indefinable woe had passed over. They did not care now to look at the final phase of the eclipse, as the darkness wore off; they had beheld the crowning spectacle ; they would not weaken the impression by looking at the partial obscuration; and soon the whole crowd melted away—leaving the astronomers to continue their observations alone."

Perhaps the most attractive chapter in the book is the last, in which the question of the plurality of worlds is discussed at considerable length and with great ability. Principal Leitch reasons the question with far more caution than we should have expected from the animation with which he expatiates on the other subjects on which he treats, and states very ably all the conspicuous arguments pro and con. His own opinion on the question is that many of the planets forming part of our solar system are not yet in that normal condition from which we can, with any degree of probability, infer that they are inhabited by living beings. Others, however, such as Mars, Jupiter, and Saturn, may not vary so much from the conditions under which we find life existing on our planet, as would warrant us in concluding with Professor Whewell, that they are uninhabitable. We suspect that something more might be said than this, in perfect accordance with the arguments based on the conditions of existing life. It cannot be denied, for example, that it is in strict accordance with analogy that other planetary systems exist, besides that to which our world belongs; and if this is granted, it would be arguing, not on a probability, but on the very highest improbability, that many planets belonging to these systems have not all the conditions for supporting life, and high intellectual life, such as our world contains. Of course we cannot be certain of the existence of other planetary systems besides our own; but we are certain of the existence of thousands of other suns, and there is nothing more reasonable than the supposition that they are the centres from whence ray forth to attendant planets the same power as that which is dispensed from our own luminary. And we suspect that from the vast number of these planetary centres the law of chances itself, to sink altogether the analogical argument, would favour the idea that some of these distant planets have at any rate conditions as favourable to the existence of animated beings as our world possesses. Principal Leitch's book, besides discussing questions like these, contains also a great deal of information in reference to recent discoveries in astronomy, is illustrated by excellent engravings (some of them —the telescopic views of the moon—being taken from photographs) and has appended to it a valuable synopsis of all the most prominent facts in the science. We heartily wish for it a large circulation.

On February 25th 1863 Leitch gave a sermon at St Andrews Church in Belleville, Ontario. Three days later his lecture on the sun was reprinted south of the border in the prestigious publication *Scientific American*. The reaction must have been favorable because six months later they also unpublished an entirely new sequel. It was called *Age of the Sun - Force and Heat*. Despite the fact that with over 150 years of hindsight we can detect all manner of errors in this essay, the editors at the time were clearly impressed. Here it is in its entirety.

The following extracts are from the Canadian Presbyterian, communicated by Principal Leitch. The subject is of general and profound interest, and it is treated with philosophical ability. Principal Leitch seems to be perfectly familiar with mechanical science:

"Perhaps the most daring attempt of astronomy in modern times is that of fixing the age of the sun as an incandescent light-giving body, and that of the earth as a solid inhabitable globe. In reference to the earth, geology plainly indicates successive periods or chapters of its history ; but no scale has been furnished of the length of the periods, and no approximation has hitherto been made to the whole period, from the first to the last page of the geological record. Science has at last attempted to assign an approximate date to the laying of the foundation stone of our world. A scale has been found by which the whole period can be measured within certain limits. You cannot, as in the section of a tree, tell to what year each layer belongs; but you can assign a date within limits to the first page in the record: or, in other words, to the first solidification of the earth.

"Again, as to the sun, its past physical history seemed to be entirely withdrawn even from speculation. He has enlightened our globe from one generation to another without any apparent diminution of strength, and we have formed the instinctive belief that no limit in the past or any in the future can be assigned to his functions. No proof of progress or decay has been detected; and it has been thought that nothing but the fiat of the Almighty can quench his rays. Principles have now been recognized, however, which enable us to assign limits, and to show that he has not shone from a past eternity, and that he has a limited existence as an incandescent body. This limit assigned to the solar system forces us to recognize the hand of a Creator.

"In order to understand the manner in which a limit is set to the past history of the sun, it is necessary to advert to the dynamical theory of heat, which has recently been reduced to a strictly scientific form. The expression of this theory is that heat is but a form of force, and that for so much heat there is an equivalent of force, and that for a given force there is an equivalent heat. This has been acknowledged in a loose general manner. For example, the heat of the furnace gives its power to the steam-engine; and in a similar way power or energy can be converted into heat. The power of a steam-engine or a water-wheel may be employed to produce heat. Where water-power is abundant, it is employed to produce friction between iron plates, and these plates become so hot that they serve as a stove. Again, the blacksmith can convert the power of his arm into heat when he hammers a piece of iron till it is red hot, and sufficient to light his fire. Force is converted into heat when the axles of a railway car take fire. The power of your finger is converted into heat when you pull the trigger of a flint lock - The spark is the heat product of the power of your finger. The obvious relation between force and heat has always been acknowledged, but it is only recently that the exact quantitative relation has been determined. The relation is thus expressed: 'a unit of heat is equivalent to 772 foot-pounds. By a unit of heat is meant heat sufficient to raise one pail of water 1° Fah. Suppose one pound of water enclosed in a vessel fell from a height of 772 feet, it would be found that it had become warmer by 1° Fah. That is, the force of the concussion has been converted into so much heat. On the other hand, if this 1° Fah., of heat could be extracted from a pound of water and applied to move an engine, it would raise, if there was no friction or loss of power, a pound of water to a height of 772 feet. The great law of force or energy is that its sum is ever the same. It cannot be annihilated. It may change from one form to the other, but the sum is ever the same. If there is a loss in mechanical power, there is a gain in some other force, such as heat, electricity, or chemical affinity. The mechanical power of the Falls of Niagara is lost as such when it reaches the bottom, but it only changes its form, for it only becomes heat; and this heat, if all applied to an engine, would raise the whole mass again to its former level. The heat of the furnace of the steamer is converted into the mechanical power of the engine. This power is re-converted into heat by the blow of the paddle, and the impact of the ship upon the water. What is lost in one form is gained in another. The sum is always the same. It is like a sand-glass; the sand is always the same in amount, though it is constantly changing from one end to the other.

"Let us apply this principle to the heat of the sun. When a ball is discharged from a gun and strikes an object, it is found that both the ball and the object struck have risen in temperature. If the force is sufficiently great you cannot touch the ball, it is so hot; and just in proportion to the power of the gun will be the heat of the ball. If the power be sufficiently great, the heat may be so intense as to bring it to a white heat and melt the ball. The meteoric stones that sometimes fall to our earth may be regarded as balls, but moving with much greater velocity. They strike against our atmosphere with so much force that the force is converted into heat, so intense that they glow or become incandescent. Suppose our earth, in its revolution, struck against some opposing object like a target, what would be the consequence? The force would be converted into heat, and the velocity is so great —twenty miles a second—that it would be immediately brought to the melting point. It would glow like the sun, and become a luminous body. The heat would be equal to that produced by the burning of fourteen earths made of coal. But this is not all. It would then fall into the sun, and would by its loss of momentum produce a heat 400 times greater than before, and it would be seen on the sun's furnace as a bright luminous spot.

"The force of the Earth falling upon the sun would communicate a heat to the sun equivalent to the heat emitted by the sun in a century. It would serve as fuel for that length of time. Now, the heat of the sun is most probably due to this source, the conversion of power into heat. It is probable that it is not a combustion. If the sun were composed of coal, it would last at the present rate only 5,000 years. The sun, in all probability, is not a burning but an incandescent body. Its light is rather that of a glowing molten metal than that of a burning furnace. But it is impossible that the sun should constantly be giving out heat, without either losing heat or being supplied with new fuel. We know the heat of the sun. Each point is about thirty times hotter than the furnace of a locomotive, that is, a square foot of the sun's surface gives thirty times more heat than a square foot of grating in a locomotive. Yet the mass of the sun is so great that it would. require 3,500 solar systems, if made of coal, to account for the heat of the sun. Assuming that the heat of the sun has been kept up by meteoric bodies falling into it, and proof has been given of such fall, it is possible from the mass of the solar system to determine approximately the period during which the sun has shone as a luminary. On boarding a steamer you can by examining the hold for coals, and ascertaining its capacity, tell approximately how long she has been on her voyage. Limits can be set to the fuel of the solar system, and therefore limits can also be assigned to the existence of the sun as our luminary. The limits lie between 100 millions and 400 millions of years. These are enormous periods, but still they are definite. The mass is so great, and the cooling is so slow, that, even on the supposition that no fuel was added, it might be five or six thousand years before the sun cooled down a single degree."

In June 1863 in a letter to another undisclosed recipient Leitch revealed that he maintained an account with Bryson's of Edinburgh. Bryson's and Sons was a long standing business at 66 Princes Street which specialized in all manner of instruments, including watches, clocks, microscopes, thermometers, hygrometers and barometers. In this case Leitch mentioned that he was buying "slides for the astronomical lectures." Leitch also purchased optical items from "mathematical and philosophical instrument makers" Alexander Adie's & Son, which was located at 50 Princes Street. Both of these shops were just opposite the Royal Scottish Academy and are on the same street and a two minute walk from where Jules Verne had stayed in 1859.

On the 12th August 1863 Leitch became a member of the Synod court during his tour of the Maritimes. He visited New Brunswick, Nova Scotia and Prince Edward Island. This is his only known tour of the eastern provinces of Canada, where there was a large Scottish population. Some of his time there was recorded in the book *One Woman's Charlottetown Diaries of Margaret Gray Lord.*

Leitch wrote to a friend in Scotland just before his death. The letter included some of his last thoughts.

"A few weeks before his death he sent a letter to an old co-presbyter. It was signed by the Rev. Doctor's own well-known hand. We give nearly the whole of it, which we shall name The Farewell.— *"I dare say, you have been wondering why I should not have replied to your letter sooner. The employment of the pen of a kind friend will reveal to you the reason of my delay. I have been confined, with serious illness, for about two months, to my bed. I rallied for some time, but I am now weaker than ever. Though my doctor gives me hope of recovery, my tenure of life is very uncertain. I fear that I have miscalculated my own strength and the rigour of the climate, so that I have been obliged to succumb. Instead of enjoying the repose of last summer's vacation, I laboured constantly in the Lower Provinces; so that, when I returned to my winter duties, my strength was quite exhausted. I am now so entirely broken down that I am unable to do any portion of my duties. My medical man gives me good hope, that by perfect repose, in Scotland, during the next vacation, I may entirely recover. Still, my feelings of prostration tell me that I ought not to be too sanguine. If I am spared, I will return to Scotland at the earliest opportunity. Please remember me kindly to the family, and to my co-presbyters, old and young... My situation here has not been one of ease or repose, and has contrasted much with the tranquil life of my old parish. Still, I have reason to think that my labours have been in a great measure blessed. During the four years I have presided over this college, I have reason to believe that many young men have gone forth to labour in the vineyard of the Lord, in the right spirit. Most of them, I think are not inferior in talent to the students at home; and many are imbued with a devoted missionary spirit, and I believe that future generations will reap the fruits of their devotedness."*

On May 14th 1864 the *Globe and Mail* newspaper described his funeral in Kingston. An obituary confirmed that he had lectured on mathematics at the Andersonian. There was also evidence of more writings, some published, some not.

"Also he is well known to have been the author of certain articles, in which, in a masterly manner, his views of the late accomplished divine, Dr. Wardlaw of Glasgow, on the subject of miracles, he controverted. These articles created great interest at the time of their appearance, and the subject came in consequence to have special attractions for their author. In one of the latest conversations which the writer had with him, he was led to understand that the Principal had a work all but ready for the press, on the leading questions pertaining to miracles as raised and discussed in modern times."

A lengthy review of Wardlaw's book about miracles appeared anonymously in the April 1854 issue of *Kitto's Journal of Sacred Literature*. Since we know Leitch wrote for this magazine and we know he wrote on this subject, it is probable that this is the article in question.

Adverts for Leitch's book in New York's *The Nation* ran well into 1867 and were accompanied by a review which stated, *"To the publisher of this work the community is under great obligation and they must feel a worthy satisfaction in knowing that he is an instrument of so much good in thus aiding to unfold the curtain of mist that displays the divine Shechinah."*

In 1867 Strahan anthologised Sir John Herschel's work from *Good Words*. *Familiar Lectures on Scientific Subjects* had the exact same embossed gold-leaf image on the cover as Leitch's book. They make a matching pair.

Finally, the subject of whether Leitch was truly forgotten and had no influence on subsequent generations of scientists is extremely difficult to answer without concrete evidence. Shortly after his death a book entitled *Architecture of the Heavens A New Theory of the Universe and the Extent of the Deluge* by Ezekiel Wiggins was published. Wiggins disagreed with Leitch's theory of how the sun worked and attempted to discredit his opinions, but with little success. Although Leitch (along with everyone else) was wrong about the fundamental forces at work, his assessment of the inevitability of the demise of the solar system was correct, something Wiggins disputed. As has been related on page 104, Leitch's essay on space flight was republished, unattributed, by Isbister in 1877 in the juvenile volume *Half Hours in Air and Sky*. Despite this, Leitch was still cited by name in several other scientific works after his death. In 1883 in *The Heavenly Bodies* by William Miller; in *Lunar Science* by Rev Thomas Harley FRAS in 1886; in the *Journal and Proceedings of the Hamilton Scientific Association* in 1889; and in 1890 in *The Fuel of the Sun* by W. Mattieu Williams FRAS. In January 1902 Rev. D.B. Marsh of the Hamilton Scientific Association listed his choice of good books to begin studying astronomy, *Half Hours in Air and Sky* was listed with *Solar System* by Thomas Dick, *Popular Astronomy* by Camille Flammarion and *Star Maps* by Richard Proctor. You could scarcely choose to be in better company.

Of course the most important find would be any proof that there was a connection between Leitch and one of the later pioneers of space flight.

Robert Goddard grew up in Roxbury, a suburb of Boston Massachusetts. In his diary he tells how in 1898 at the age of 16 he attended the *English High School* in Boston. His biographer Milton Lehman says that during this time he was regularly borrowing books about astronomy and science from the library. He was also completely enraptured by the science fiction space flight adventures of H.G. Wells and Garrett Serviss which appeared in the Boston Post early that year.

The South-End branch of the Boston Public Library, which was closest to Goddard's family home, first opened in August of 1877, five years before Goddard was born. At the end of that same year a copy of the brand new *Half Hours in Air and Sky* was purchased by the library, but the copy was assigned to the main branch a few miles away. This seems peculiar because the South-End branch already possessed several similar books including *Half Hours with the Telescope* by Richard Proctor, published in 1873. The other volumes on astronomy at South-End were by Ormsby Mitchel, Thomas Dick and John Herschel, all of whom were connected to Leitch. In fact Mitchel's book was published just a few months after his meeting with Leitch and was about astronomy and religion. (*Astronomy and the Bible*, 1863)

In 1881, just before Goddard was born, the prestigious *English High School* moved to Montgomery Street, barely half a mile from the main branch where the book had first been sent. Then the South-End branch of the library *moved into his future school* and all of the books in the branches were shuffled around. The head of the Boston Public Library reported that during that year 1.2M books were borrowed from the library and that 900,000 of those were borrowed by students from Boston schools. Was the book still there in 1898? At that time it was unheard of for libraries to discard books without good reason; it was cheaper to repair them than replace them. In fact *Half Hours in Air and Sky* was in the Detroit public library for more than two decades until 1908, and in the Baltimore public library until at least 1890. So for about a year Robert Goddard probably studied within half a mile, of a copy of Leitch's unattributed essay on spaceflight. Perhaps it may have even been in the same building; right at the exact time that he was first getting interested in space.

Around the same time, Robert Goddard's father bought him a subscription to *Scientific American* magazine. He remained an avid reader and then a contributor to that prestigious journal for decades, and when he was ridiculed for suggesting a rocket would work in space, he rebutted the naysayers in 1921...in *Scientific American*.

Further reading:

The Globe Toronto Jul 11 1844, Oct 8 1844 , Sep 29 1847, Dec 29 1847, Apr 14 1849, Dec 13 1859, May 7 1860, Nov 10 1860, Jan 20 1862 , Apr 30 1862, May 29 1862 , Jun 7 1862, Jun 26 1862, Dec 6 1862, Mar 6 1863, May 14 1863, Nov 18 1863, Feb 13 1864, May 12 1864, May 14 1864

Queen's University, Roger Graham and Frederick Gibson, McGill Queens University Press, 1978

Queen's University at Kingston; the first century of a Scottish-Canadian foundation, 1841-1941. Delano Dexter Calvin, The Trustees of the University, 1941

A disciplined intelligence: critical inquiry and Canadian thought in the Victorian era, A.B. McKillop, McGill Queens University Press, 2001

Daldy-Isbister Catalog, 1876

The Extra-terrestrial Life Debate by Michael Crow, Cambridge University Press, 1986

The Missionary's Warrant by William Leitch,The Church of Scotland Pulpit, Macphail, Edinburgh, 1845 The American Golfer, Volume 16. Story on Cecilia Leitch

Astronomy a Century Ago by A.V. Douglas, Journal of the Royal Astronomical Society of Canada, Vol. 58, 1964

Reflections on the Astronomy of Glasgow, David Clarke, Edinburgh University Press, 2013

MacPhail's Edinburgh Ecclesiastical Review, Myles MacPhail, Edinburgh 1848-1851

The Presbyterian, The Presbyterian Church of Canada, 1859-1864

Acknowledgments:

With thanks to the staff at the University of Guelph Library Annex, McMaster University Archives, Knox College and Trinity College Toronto, Mr. Frank Winter, Mr. Hugh Laidlaw at Monimail, Mr. Michael Ciancone, Mr David Baker, Ms. Julie Greenhill at St Andrews University, Mr. Hugh Chambers, Mrs. Patricia Godwin, Mr. Hugh Black, Mr. Chuck Black, Mr. John Pearson at the Women's Golfing Museum, Ms. Jessica Hood at Cataraqui Cemetery, Principal Daniel Woolf, Ms. Mary-Rose Lalande, Ms. Deirdre Bryden, Dr. Alvan Bregman, Mr. Andrew Carroll, Professor Duncan McDowall and Dr. Paul Banfield at Queen's University, Rachael Egan at Glasgow University, Mr. Christopher Markwell and finally Mr Gordon Cameron of Stuart & Stuart.

The Nova Scotian. The steamer which carried William Leitch to Canada in October 1860

About the Author:

Robert Godwin is the author and editor of dozens of books on spaceflight. He is a member of the International Astronautical Federation History Committee and the American Astronautical Society History Committee. He was Space Curator at the Canadian Air & Space Museum in Toronto. He has appeared on the BBC, CBC, Discovery Channel, History Channel, CTV and dozens of syndicated television and radio shows. He was a consultant on the award-winning television series *Rocket Science, Legends* and *Mars Rising*. His imprint *Apogee Books* was winner of the "Best Presentation of Space" award from the Space Frontier Foundation. The Minor Planet *Godwin 4252* is named after him and his brother for their contributions in documenting space history. He lives in Burlington Ontario.

Robert Godwin and Canadian astronaut Chris Hadfield examine first editions of *God's Glory in the Heavens* and *Good Words.*

Other books by the same author:

The NASA Mission Reports by Robert Godwin, Apogee Books, 1998-2016

Dynasoar Hypersonic Strategic Weapons System by Robert Godwin, Apogee Books, 2003

Apollo Project The Test Program, by Robert Godwin, Apogee Books, 2005

Apollo 11 First Men on the Moon, by Robert Godwin, Apogee Books, 2005

Mars by Robert Godwin, Apogee Books, 2005

Russian Spacecraft, by Robert Godwin, Apogee Books, 2006

Apollo Project Exploring the Moon, by Robert Godwin, Apogee Books, 2006

The Lunar Exploration Scrapbook, by Robert Godwin, Apogee Books 2007

Around the World in 65 Days - The Journal of the Real Phileas Fogg, by George Griffith and Robert Godwin, Apogee Books, 2010

New Horizons, by Robert Godwin, Apogee Prime, 2011

Arthur C. Clarke A Life Remembered, by Fred Clarke, Mark Stewart, Kelvin Long and Robert Godwin, Apogee Prime 2012

Space Shuttle Fact Archive, by Robert Godwin, Apogee Books 2015

2001 The Heritage & Legacy of the Space Odyssey by Frederick I. Ordway III and Robert Godwin, Apogee Prime, 2015

www.apogeebooks.com

Index

solar power 166

spin-stabilization 61

Sputnik 6, 102, 138

SS Great Eastern 73, 87

Stars and the Earth 146

Stern, Alan 79

Strada 149

Strahan, Alexander 60, 68, 69, 70, 71, 74, 82, 83, 92, 99, 103, 104, 105, 106, 114, 124, 142, 153, 154, 181

Sun, (Lecture on the) 160

T

Tait, Peter Guthrie 79

telegraph 77, 149

Tennyson, Lord 103

Tesla, Nikola 83

The Claddy 21

Theology and Science 144

third law (Newton's) 12, 14, 97

Thomson, William 24, 27, 28, 29, 34, 61, 76, 79, 82, 134

Trinity College 45, 144

Tsiolkovsky, Konstantin 5, 11, 12, 13, 14, 19, 22, 23, 102, 137, 139

U

USS San Jacinto 116

V

Valier, Max 85

Verne, Jules 5, 6, 9, 10, 11, 14, 16, 18, 19, 22, 63, 64, 65, 73, 105, 120, 137, 139, 179

von Braun, Wernher 18

Voyager 2 108

Vulcan 70, 78, 79

W

Watt Institution 55

Watt, James 23, 24, 154

Wells, H.G. 82, 182

Whewell, William 23, 44, 45, 46, 47, 48, 49, 50, 127, 144

Winter in Canada (A) 68, 121, 124

Winter, Frank 6, 144

Wollaston, William 23

Wollstonecraft, Mary 121

Woolwich Arsenal 101, 119

Wordsworth, William 30, 74

Wright Brothers 99

X

X-15 aircraft 140

The **Baird of Bute Society** (a not-for-profit originally established in Scotland to recognise the pioneering work of aviator Andrew Blain Baird) working in conjunction with the University of Strathclyde Glasgow, have established a scholarship:

"This new Baird of Bute Leitch Space Scholarship recognises Dr William Leitch, another inspiring Scot from the Isle of Bute, who in the 1860's in his scientific writings proposed that propulsion in space was best achieved by the use of rocket power - many decades ahead of Goddard and the others previously credited.

"The Society was very pleased to have Brooklyn Adkins as the first recipient of this exciting new Scholarship and on September 9th 2016 recognised her at their annual Awards Dinner on the Isle of Bute hosted by its Patron Lord Smith of Kelvin."

Queen's University, Wallace Hall January 11th 2017, under the image of Principal Leitch on the occasion of presenting the Baird of Bute award to Principal Leitch. (l to r) Dr Alvan Bregman, Prof. Duncan McDowall, Mr. Christopher Markwell, Principal Daniel Woolf, Mr. Robert Godwin, Dr. Paul Banfield (image courtesy of Queens University)